121 Advances in Polymer Science

Polymer Synthesis/ Polymer Engineering

With contributions by
K. Ganesh, K. E. Geckeler, K. Kishore,
S. Kobayashi, D. B. Priddy, S. Shoda,
H. Uyama

With 48 Figures and 23 Tables

Springer-Verlag Berlin Heidelberg GmbH

ISBN 978-3-662-14871-6   ISBN 978-3-540-49060-9 (eBook)
DOI 10.1007/978-3-540-49060-9

© Springer-Verlag Berlin Heidelberg 1995
Originally published by Springer-Verlag Berlin Heidelberg New York in 1995.
Softcover reprint of the hardcover 1st edition 1995
Library of Congress Catalog Card Number 61-642

Typesetting: Macmillan India Ltd., Bangalore-25
SPIN: 10470582   02/3020 - 5 4 3 2 1 0 - Printed on acid-free paper

Editors

Table of Contents

Enzymatic Polymerization and Oligomerization

Shiro Kobayashi, Shin-ichiro Shoda, and Hiroshi Uyama
Department of Molecular Chemistry and Engineering, Tohoku University,
Sendai, 980 Japan

Polymerizations and oligomerizations catalyzed by various enzymes including hydrolases, transferases and peroxidases are reviewed. Optically active polyesters have efficiently been prepared by lipase-catalyzed asymmetric polymerizations using various dicarboxylic acid derivatives as monomers. A novel method for regio- and stereoselective synthesis of polysaccharides and oligosaccharides has been developed by enzymatic polycondensations of glycosyl fluoride monomers. The methodology has successfully been applied to the first in vitro synthesis of cellulose via a nonbiosynthetic path. A variety of polyaniline derivatives have been produced by peroxidase-catalyzed oxidation polymerizations of aniline derivatives in relation to their electronic properties.

1 Introduction

The progress of macromolecular science has been strongly attributed to the creation of a new class of polymers which are produced by the chemical modification of naturally occurring biopolymers or by the polymerization of various monomers having novel structures. All of the polymerization reactions utilized in polymer synthesis can be classified into the following four categories according to the chemical nature of the polymerization catalysts: radical polymerization [1], cationic polymerization [2], anionic polymerization [3], and coordination polymerization [4]. As the structures of targeting polymers are becoming more complicated, polymerization catalysts to promote the reaction under milder conditions and with higher regio- and stereoselectivities are required, and enormous efforts have been devoted to the development of new polymerization catalysts in order to achieve these requirements.

Enzymes have several remarkable catalytic properties compared with other types of catalyst in terms of the selectivity, high catalytic activity, lack of undesirable side-reactions, and operation under mild conditions. In addition, with the advent of genetic engineering, it will be possible to produce a wider range of enzymes on a larger scale expanding the number of enzymes available for synthetic reactions. Although many synthetic reactions catalyzed by enzymes have appeared in synthetic organic chemistry [5–10], few examples on polymerization reactions have been reported so far. This is probably due to the fact that the structural variation of our synthetic targets has only in recent years begun to make highly selective polymerizations necessary in response to the increasing demands for the production of various functional polymers in material science. The present review deals with recent advances in polymerization reactions catalyzed by an enzyme, proposing the term "enzymatic polymerization" as a novel concept in polymer synthesis.

In nature, all of the reactions producing important biopolymers are catalyzed by enzymes. It is, therefore, necessary to define clearly the term "enzymatic polymerization". There are generally three classes of biopolymer syntheses catalyzed by an enzyme.

1. Biosynthesis in vivo (in living cells) via biosynthetic pathways, e.g. naturally occurring reactions in all living systems.

2. Biosynthesis in vitro (outside cells) via biosynthetic pathways, e.g. a polymerization of a substrate of a phosphate derivative catalyzed by a polymerase enzyme in the cell-free extract.

3. Chemical synthesis in vitro (in test tubes) via non-biosynthetic pathways catalyzed by an isolated enzyme.

Now, we can define the above class (3) as "enzymatic polymerization". Biosyntheses via classes (1) and (2) produce naturally occurring biopolymers (macromolecules) in almost all cases. "Enzymatic polymerizations" of class (3), on the other hand, allow us to produce not only naturally occurring biopolymers but non-natural synthetic polymers, depending on the combination of substrate monomers and enzymes. The first chapter in this article deals with a brief review of in vivo polymerization reactions, namely biosynthesis of nucleic acids, proteins, polysaccharides, rubbers, and microbial polyesters, which belong to classes (1) and (2). The following chapters are concerned with the recently developed "enzymatic polymerization" catalyzed by hydrolases, transferases, and peroxidases, which belong to class (3).

2 A Brief Review of Polymerization Reactions via Biosynthetic Pathways

Biologically important macromolecules such as nucleic acids, proteins, and polysaccharides are not formed by direct condensation of the corresponding constructing units, nucleotides, amino acids, and monosaccharides, respectively. The activated precursors of phosphate derivatives formed by utilizing a high energy compound, namely adenosine triphosphate (ATP), are key monomeric substances for the synthesis of these macromolecules. The biosynthetic process therefore consists of the following two steps. The first step involves the formation of an activated monomer of phosphate ester derivative. The second step is the polymerization process catalyzed by an enzyme having high stereospecificity toward the substrate.

2.1 Nucleic Acids

A typical example of enzymatic synthesis of nucleic acids is the transcription of the genetic code from DNA to messenger RNA (mRNA) [11]. The polymerization is catalyzed by RNA polymerase which exists in virtually all cells. This enzyme connects ribonucleotides by catalyzing the formation of the internucleotide 3'-5' phosphodiester bonds. The monomers are triphosphates of adenine (A), cytosine (C), guanine (G), and uracil (U), whose nucleic acid base parts specifically pair with the corresponding bases of DNA, thymine (T), guanine (G), cytosine (C), and adenine (A), respectively. The polymerization takes place by releasing pyrophosphate (Fig. 1).

2.2 Proteins

Protein synthesis begins in the cell nucleus with synthesis of mRNA. It acts as a template on the surface of the ribosomes by making a triplet pair with transfer RNA (tRNA) which carries amino acid residues, to the site of mRNA. Codon-anticodon pairing occurs between mRNA and tRNA, and an amide bond

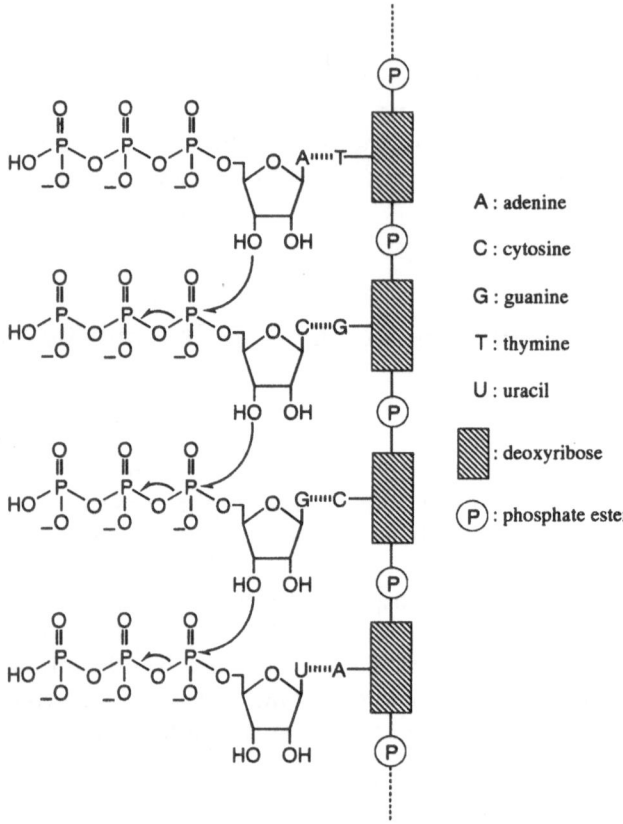

A : adenine

C : cytosine

G : guanine

T : thymine

U : uracil

: deoxyribose

(P) : phosphate ester

Fig. 1. Transcription of genetic code from DNA to messenger RNA (mRNA)

formation is achieved enzymatically (Fig. 2). After the first amide bond is formed the ribosome moves to the next codon on mRNA [12].

Concerning the activation of amino acids for constructing the peptide bond, the following process has been elucidated (Fig. 3). The first step is the formation of adenosine monophosphate of an amino acid (aminoacyl-AMP) by the reaction of an amino acid and adenosine triphosphate (ATP) catalyzed by aminoacyl-tRNA synthase (ARS). The resulting aminoacyl-AMP is further attacked by a hydroxyl group of a specific transfer RNA giving rise to an aminoacyl-tRNA as a precursor for the peptide bond formation.

2.3 Polysaccharides

The typical naturally occurring polysaccharide, cellulose, is enzymatically synthesized in vivo starting from an activated form of glucose, uridine diphosphate

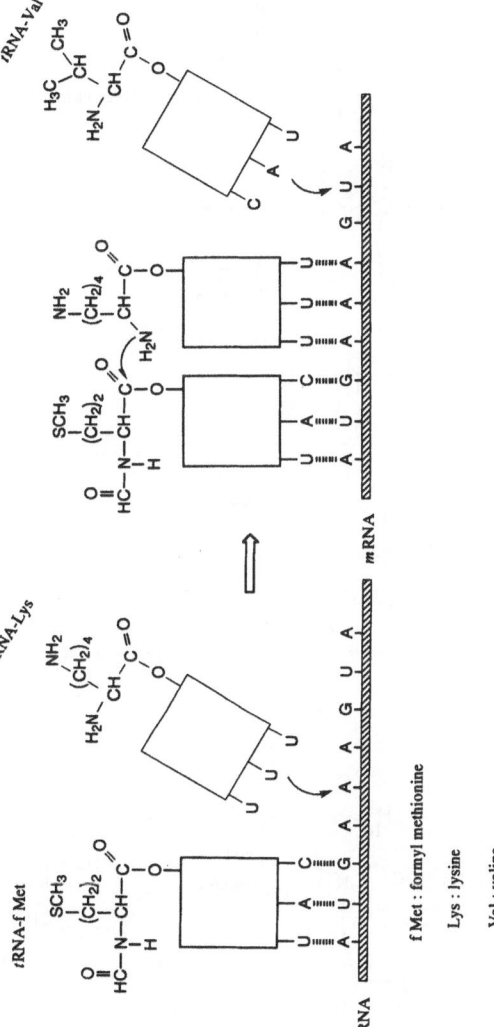

Fig. 2. Peptide bond formation via codon-anticodon pairing between mRNA and tRNA on ribosome

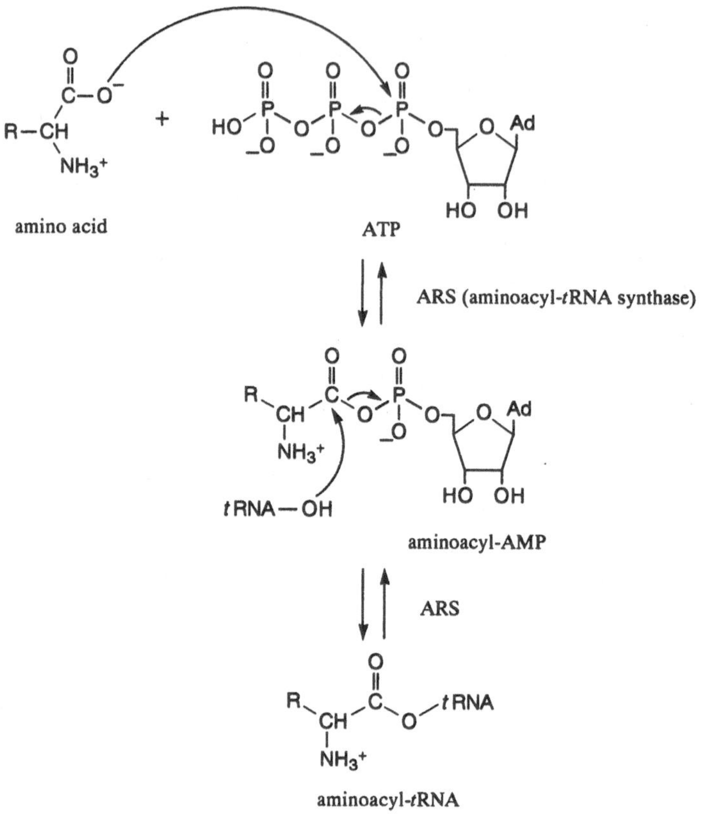

Fig. 3. Activation process of amino acid: Formation of aminoacyl-tRNA

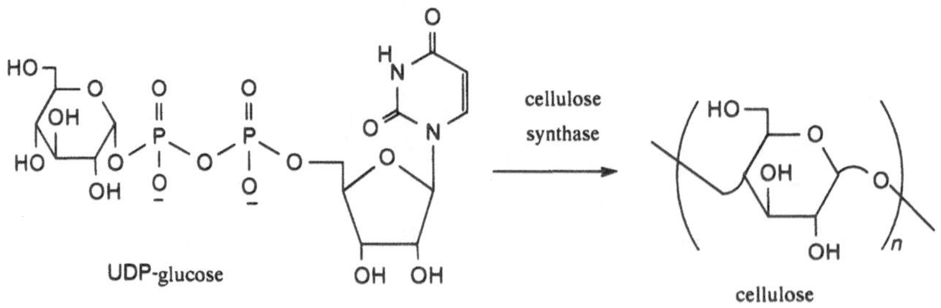

glucose (UDP-glucose) catalyzed by cellulose synthase. Formation of cellulose according to a biosynthetic pathway using *Acetobacter xylinum* [13, 14] or *Phaseolus aureus* extracts [15] with a nucleoside diphosphate sugar (ADP-, CDP- GDP-, or UDP-glucose) as substrate has been reported.

2.4 Rubbers

Natural rubber can be regarded as a 1,4-addition polymer of isoprene. The basic building block of five carbons for the polymerization is 3-isopentenylpyrophosphate (3-IPP). The first reaction is an enzymatic isomerization of the olefin of 3-IPP to 2-isopentenylpyrophosphate (2-IPP). The carbon-carbon bond formation between these two pyrophosphates initiates the polymerization in which the pyrophosphate group acts as a leaving group. The isoprene units of natural rubber are all linked in a head-to-tail fashion and all of the double bonds have a *cis*-structure. The stereocontrol of the formation of the *cis*-unit is achieved by the function of the elongation factor which combines with the farnesyl pyrophosphate (FPP) synthase [16].

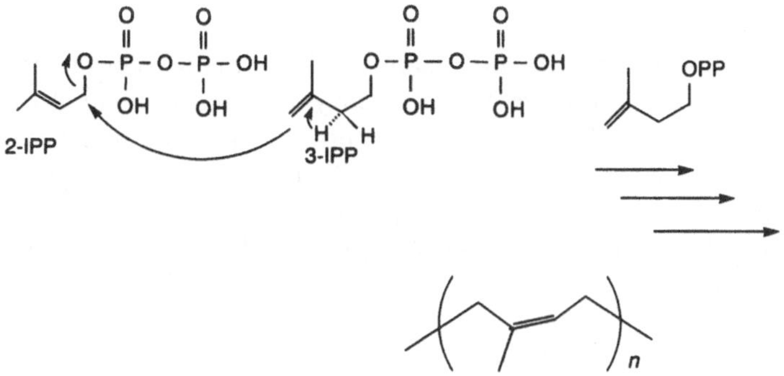

2.5 Polyester Synthesis by Microorganisms

The chemistry of poly (β-hydroxybutyrate) (PHB) involving structure, biosynthesis, and degradation has attracted the attention of scientists in connection with the environmental problems. In 1985, Holmes et al. at ICI published that a

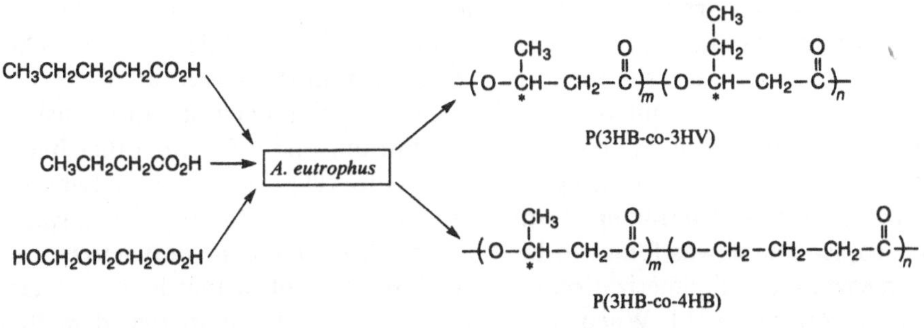

Fig. 4. Synthesis of random copolyester by *Alcaligenes eutrophus*

random copolymer P(3HB-*co*-3HV) of 3-hydroxybutyrate (3HB) and 3-hydroxyvalerate (3HV) is produced by *Alcaligenes eutrophus* growing on glucose and propionic acid [17]. The ratio of 3HV can widely be controlled from 0 to 95 mol % by growing the microorganism in the presence of valeric acid and butyric acid [18]. The addition of 4-hydroxybutyric acid gave a new random copolyester having a 3HB unit and a 4-hydroxybutyrate (4HB) unit [19] (Fig. 4). Studies on the degradability of these poly(hydroxyalkanoates) by the action of a microorganism from the soil is extensively in progress in order to create new biodegradable polymers [20].

3 Polyester Synthesis by Enzyme Catalysts

Since Klibanov [21] demonstrated the first example of enzymatic esterification and transesterification in organic solvents, many enzymatic reactions in anhydrous media have been reported. It is now well known that some hydrolytic enzymes are stable even in organic solvents and can be used for certain types of condensation reactions that are difficult or impossible to achieve in aqueous media [22–28]. Most of these reactions concern lipase-catalyzed esterifications and transesterifications [29].

3.1 Polymerization of Achiral Carboxylic Acid Derivatives

The first paper describing an enzymatic synthesis of polyesters appeared in 1986, where achiral hydroxy acids were used as monomers [30]. The polycondensation of 12-hydroxyoctadecanoic acid **1a**, 12-hydroxy-*cis*-9-octadecenoic acid **1b**, 16-hydroxyhexadecanoic acid **1c** and 12-hydroxydodecanoic acid **1d** catalyzed by *Candida rugosa* lipase and *Chromobacterium viscosum* lipase afforded the corresponding polyesters **2** having a long alkyl chain. The reaction was carried out in water or in a non-polar organic solvent, and the molecular weight (M_n) of the resulting polymers was 600–1300. Monomers **1a** and **1b** bearing a secondary hydroxy group quickly polymerize both in water and in isooctane to yield oligomers having wider molecular weight distributions (Mn = 1000–1300). In the reaction process, the monomer concentration decreases rapidly before any high molecular weight polymer is formed. In this reaction, an emulsifier plays an important role in shifting the equilibrium point. On the other hand, hydroxy acids, **1c** and **1d**, bearing a primary hydroxyl group slowly polymerize to yield trimers and tetramers as the main products. Of the organic solvents screened, isooctane was found to be the most effective for the polymerization.

An enzymatic oligomerization versus lactonization of various hydroxyesters has been reported [31]. When unsubstituted β, δ, and ε-hydroxyacid methyl esters **3** are subjected to the action of Porcine pancreatic lipase in anhydrous

$$n \quad \underset{\mathbf{1}}{HO-\overset{\overset{\displaystyle R^1}{|}}{C}H-R^2-\overset{\overset{\displaystyle O}{\|}}{C}-OH} \longrightarrow H\left(\!O\overset{\overset{\displaystyle R^1}{|}}{C}H-R^2-\overset{\overset{\displaystyle O}{\|}}{C}\!\right)_{\!\!n}\!OH$$

$$\underset{\mathbf{2}}{}$$

1a : $R^1 = (CH_2)_5 CH_3$, $\quad R^2 = -(CH_2)_{\overline{10}}$

1b : $R^1 = (CH_2)_5 CH_3$, $\quad R^2 = -CH_2-CH \overset{cis}{=} CH -(CH_2)_{\overline{7}}$

1c : $R^1 = H$, $\quad R^2 = -(CH_2)_{\overline{14}}$

1d : $R^1 = H$, $\quad R^2 = -(CH_2)_{\overline{10}}$

$$CH_3 - O - \overset{\overset{\displaystyle O}{\|}}{C} - (CH_2)_{\overline{m}} \ CH_2 - OH \longrightarrow$$

$$\mathbf{3}$$

$$\begin{array}{c} (CH_2)_{\overline{m}} - C \overset{\displaystyle O}{\underset{\displaystyle |}{\big\|}} \\ | \qquad | \\ CH_2 \;-\; O \\ \mathbf{4} \end{array}$$

$$\longrightarrow CH_3 - O\left[\overset{\overset{\displaystyle O}{\|}}{C} - (CH_2)_{\overline{m}} - CH_2 - O\right]_{\!n}$$

$$\mathbf{5}$$

3a : $m = 1$

3b : $m = 3$

3c : $m = 4$

organic solvents, they exclusively undergo intermolecular transesterification to afford the corresponding oligomers **5**; lactonized product **4** was not obtained. In contrast, the substituted δ-methyl-δ-hydroxyester undergoes lactonization even in concentrated solutions.

Lipase-catalyzed polymerization between an achiral dicarboxylic acid and a primary glycol in an aqueous-organic mixture was reported [32]. A poly-condensation of achiral 10-hydroxydecanoic acid catalyzed by lipase in benzene has also been claimed, however no convincing evidence was presented [33].

Enzymatic polycondensation of a dicarboxylic acid and a diol has been reported using lipase from *Candida cylindracea*, *Pseudomonas* sp., and Porcine pancreas [34]. Competition between linear polyester and macrocyclic lactone formation was observed depending on the reaction temperature.

3.2 Asymmetric Polymerization of Carboxylic Acid Derivatives

A lipase-catalyzed asymmetric polycondensation have been achieved using the racemic diester **6** and the achiral diol **7** (or the achiral diester and the racemic diol) in an organic solvent [35]. Optically active trimers and pentamers **8** were

$$ClCH_2CH_2O-\overset{\overset{O}{\|}}{C}-\overset{\overset{H}{|}}{\underset{\underset{Br}{|}}{C}}-CH_2CH_2-\overset{\overset{H}{|}}{\underset{\underset{Br}{|}}{C}}-\overset{\overset{O}{\|}}{C}OCH_2CH_2Cl \quad + \quad HO-(CH_2)_6-OH$$

$$\mathbf{6} \qquad\qquad\qquad\qquad\qquad\qquad\qquad \mathbf{7}$$

$$\xrightarrow{\text{lipase}} \left[-\overset{\overset{O}{\|}}{C}-\overset{\overset{H}{|}}{\underset{\underset{Br}{|}}{C}}-CH_2CH_2-\overset{\overset{H}{|}}{\underset{\underset{Br}{|}}{C}}-\overset{\overset{O}{\|}}{C}-O-(CH_2)_6-O- \right]_n$$

$$\mathbf{8}$$

$$Cl_3CCH_2O-\overset{\overset{O}{\|}}{C}-CH_2-\overset{\overset{CH_3}{|}}{\underset{\underset{H}{|}}{C}}-CH_2CH_2-\overset{\overset{O}{\|}}{C}-OCH_2-CCl_3 \quad + \quad HO-(CH_2)_6-OH$$

$$\mathbf{9} \qquad\qquad\qquad\qquad\qquad\qquad\qquad \mathbf{7}$$

$$\xrightarrow{\text{lipase}} \left[-\overset{\overset{O}{\|}}{C}-CH_2-\overset{\overset{CH_3}{|}}{\underset{\underset{H}{|}}{C}}-CH_2CH_2-\overset{\overset{O}{\|}}{C}-O-(CH_2)_6-O- \right]_n$$

$$\mathbf{10}$$

produced. An asymmetric enzymatic oligomerization was also observed when bis (2, 2, 2-trichloroethyl) (±)-3-methyladipate **9**, was used where the asymmetric carbon is farther from the reaction site than in **6**.

An enantioconvergent polymerization of symmetrical hydroxy diester **11** has been developed [36]. The polymerization utilizes the prochiral stereospecificity of enzymes, i.e. their ability to discriminate between enantiotopic groups of a prochiral molecule, in organic solvents. This approach is significantly potential for the asymmetric synthesis of chiral polyesters from achiral monomers possessing σ-symmetry. The enantiomer excess (ee) values (30–37%) of the resulting oligomer **12** are considerably higher than the reported values which had been obtained by a chemical polymerization of prochiral monomers with asymmetric catalysts [37].

$$HO-CH\overset{\displaystyle CH_2\text{-}CO_2CH_3}{\underset{\displaystyle CH_2\text{-}CO_2CH_3}{\big\langle}} \qquad\longrightarrow\qquad H\left[O-CH-CH_2-\overset{\overset{O}{\|}}{C}\right]_n O-CH\overset{\displaystyle CH_2\text{-}CO_2CH_3}{\underset{\displaystyle CH_2CO_2CH_3}{\big\langle}}$$

$$\mathbf{11} \qquad\qquad\qquad\qquad\qquad \underset{\underset{CO_2CH_3}{|}}{\overset{\overset{CH_2}{|}}{}} $$

$$\mathbf{12}$$

Synthesis of an optically active epoxy-substituted polyester **15** by lipase-catalyzed polymerization has been demonstrated [38]. A highly enantioselective polymerization of a chiral, epoxide-substituted diester **13** with 1,4-butanediol **14** was performed using Porcine pancreatic lipase (PPL) as catalyst. The molar ratio of the diester to the diol was adjusted to 2:1 assuming that only one enantiomer of the diester would react. The unchanged (+)-monomer (+)13

was shown to have an enantiomeric purity of > 95% by ^{1}H-NMR spectroscopy in the presence of a chiral shift reagent. Direct determination of the stereochemical purity of the polymer using the shift reagent was unsuccessful and a method for determining the optical purity of the copolymer remains to be developed.

Transesterification of diphenyl carbonates with glycols proceeds using *Candida cylindracea* lipase of Porcine liver esterase as catalyst to give the corresponding polycarbonates [39]. Enantiomeric selectivity in transesterification of diphenyl carbonate with racemic 2-butanol was observed (> 80% ee). Transesterification activity of diphenyl carbonate in a water-saturated organic solvent is strongly affected by the choice of solvent. The stability of the enzyme in dry conditions was also investigated. It is speculated that a small amount of water remains tightly bonded to the enzyme allowing retention of native conformation and hence the enzyme activity is held.

3.3 Sugar Containing Polyester

Another advantageous feature of enzymes in an esterification reaction lies in their ability to distinguish a specific hydroxyl group of a complex polyalcohol like sugar derivatives. Klibanov et al. found that *Bacillus subtilis* protease (subtilisin) catalyzes the regioselective acylation of disaccharides 16, as well as nucleosides and related compounds in N,N-dimethylformamide [40].

Sucrose-containing linear polyesters 19 have been prepared using various enzymes from sucrose 17 and dicarboxylic acid diesters 18 [41]. The polycondensation reaction proceeds in anhydrous pyridine. An alkaline protease from a *Bacillus* sp. catalyzes the esterification of sucrose with bis(2,2,2-trifluoroethyladipate) to give a sucrose-containing polyester with $M_w = 2100$. The sucrose polyester is highly water soluble and soluble in polar organic solvents.

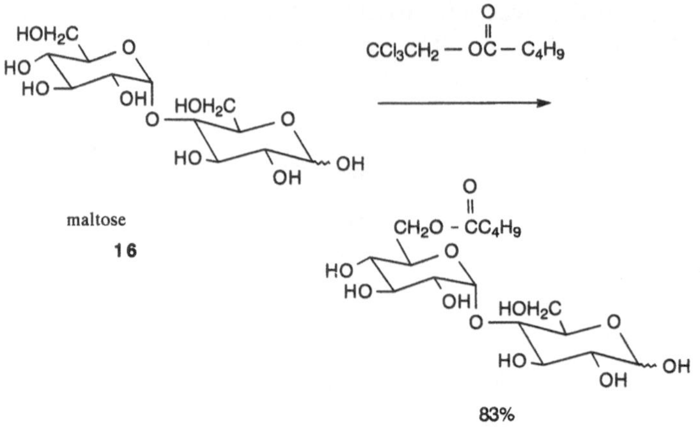

$$CCl_3CH_2 - O\overset{O}{\overset{\|}{C}} - C_4H_9$$

maltose

16

$$CH_2O - C\overset{O}{\overset{\|}{C}}C_4H_9$$

83%

>95% 6'-O-monobutyrylmaltose

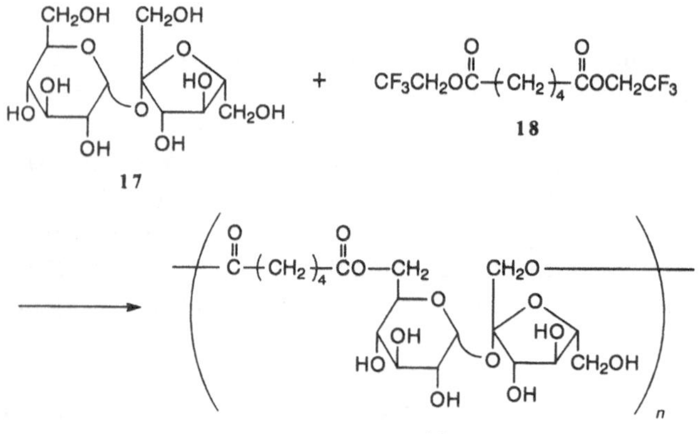

$$CF_3CH_2O\overset{O}{\overset{\|}{C}}-(CH_2)_4-\overset{O}{\overset{\|}{C}}OCH_2CF_3$$

18

17

$$\left(-\overset{O}{\overset{\|}{C}}-(CH_2)_4-\overset{O}{\overset{\|}{C}}O-CH_2 \quad CH_2O\right)_n$$

19

The polyester is degraded by the same enzyme in an aqueous solution (pH 7) to an M_w of c.a. 900 after 9 days. This polymer may have applications as a water-absorbent, biodegradable plastic.

4 Enzymatic Synthesis of Polypeptides and Related Polymers

Polypeptides are compounds of two or more amino acids having peptide bonds. Many-biologically important biopolymers belong to polypeptides. Enzymes, antibodies, hormones, storage proteins (egg white), structural proteins (keratin, collagen) have long been among the most widely investigated polymers in biological chemistry. The most powerful synthetic tool in peptide chemistry is chemical formation of peptide bonds by the use of various condensation reagents. Numerous studies have been concerned with new peptide-forming methodology [42]. The principle of a peptide bond formation is very straight-forward since an amino acid normally has only two functional groups, a carboxylic group and an amino group, causing the chemical synthesis to be the most popular method for construction of a peptide bond.

Peptide bond formation between two different amino acids can be achieved by condensing an amino acid having a free amino group with an amino acid having a free carboxyl group. The protection of the amino group and carboxyl group is an important factor in peptide synthesis, and various protecting groups have been developed. Concerning the activation of the carboxyl group, there have been reported many methodologies. Removal of the protecting groups affords the corresponding peptides having a free carboxylic acid or free amine group, which are further utilized for the next peptide forming reaction. This

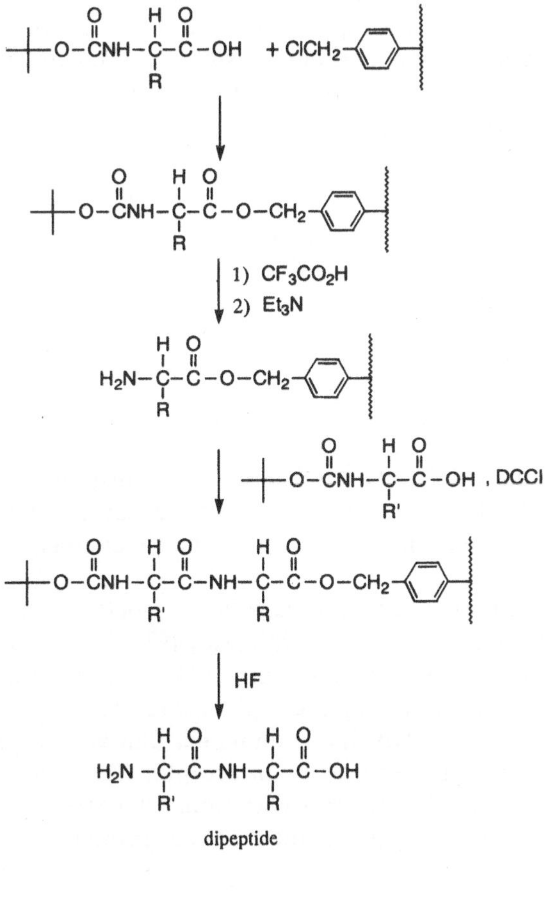

Fig. 5. Principle of peptide bond formation by solid-phase synthesis according to Merrifield

fundamental principle has been applied to solid-phase synthesis (Fig. 5). The Merrifield solid-phase peptide synthesis involves the attachment of first *N*-protected amino acid, deprotection, chain extension, repetition of these processes, and cleavage from the polymer support [43, 44].

Most of the peptide syntheses catalyzed by enzymes concern the coupling of two different amino acids or two different oligo-peptides which are important building blocks of certain biologically active polypeptides, and few reports on the polymerization of an amino acid derivative have appeared so far. The history of enzymatic peptide bond formation has been reviewed precisely in the literature [45, 46]. In these reactions, a reversed reaction of proteases was utilized for the peptide bond formation.

An enzyme-assisted polymerization of chiral fluorinated materials in organic media has been achieved [47]. Chiral fluorinated materials having two functional groups (OH or NH_2 and COOH) in the molecule **20** were polymerized with modified cellulase in organic solvents to give chiral fluorinated polymers **21** of narrow molecular weight distributions. Immobilized cellulase prepared from 2-(trifluoromethyl)propenoic acid chloride and cellulase (*Trichoderma viride*) in $CF_2ClCFCl_2$ was used as catalyst. To a suspension of the enzyme catalyst was added a chiral amide and the resulting mixture was stirred for 15 min at 40–41 °C. After 3 days of stirring at the same temperature, the resulting polymer was isolated and purified by column chromatography on silica gel. The immobilized catalyst was recovered and was available for reuse. The attachment of the enzyme to the support may stabilize the enzyme.

Polymerization of phenylalanine catalyzed by α-chymotrypsin-poly(ethylene glycol) complex was examined [48]. The solvent used is chloroform saturated by Tris buffer (pH 7). The monomer conversion was 72% giving the dimer and then the hexamer.

A new approach for the synthesis of oligopeptides **23** (dimer to octamer) has been developed using α-chymotrypsin as an enzyme catalyst [49]. The polymerization was carried out in a highly alkaline solution – pH 10–11. In this reaction, the enzyme catalyzes the successive coupling of phenylalanine from the corresponding ethyl ester (Phe-OEt) **22** in which one monomer unit acts as an acyl donor, while the other Phe-OEt acts as a nucleophilic acceptor. The yield of product polymers was dependent on pH, ester concentration, ester structure, concentration of the enzyme, and ionic strength of the reaction medium.

5 Synthesis of Polysaccharides and Oligosaccharides Using an Enzyme

Polysaccharides are among the most important biopolymers as are proteins and nucleic acids in nature. Polysaccharides are high molecular weight carbohydrates formed as a result of condensation reactions with elimination of water between the hydroxyl group at C-1 carbon atom of a monosaccharide unit and a hydroxyl group of another sugar unit. The role of the various polysaccharides and oligosaccharides, in which they are closely related to specific transformation of the information on the cell surface, has accelerated fundamental research into polysaccharides [50–53]. This situation prompted the investigation of the synthesis of various poly- and oligo-saccharides. Recent developments of new synthetic methodologies has strongly enhanced the production of various useful carbohydrates in material science, pharmaceutical science, and the food industry [54–56]. Chemical approaches, however, require complicated procedures including a regioselective blocking and deblocking of a hydroxy group in the sugar moieties in order to achieve regioselectivity. In addition, complete stereocontrol of the glycoside bond-forming reactions has not been achieved in spite of progress on the polycondensation reactions of partially protected sugar derivatives [57], and ring-opening polymerizations of anhydrosugars [58–60].

The use of an enzyme for the glycosylation process, therefore, is considered to be one of the most promising methodologies for selective construction of a glycosidic linkage. The enzymes which have been so far been utilized for glycosidic bond formation are, to our knowledge, restricted to glycosidases, glycosyl transferases, and phosphorylases. Enzymatic formation of glycosidic bond is realized by combined use of a glycosyl donor **24** and a glycosyl acceptor **26**. The former is activated by an enzyme to give a glycosyl-enzyme intermediate **25** which can be attacked by a hydroxyl group of the acceptor forming a glycosidic bond between the donor and the acceptor.

5.1 Free Sugars as Glycosyl Donors

The fundamental principle of glycosidic bond formation using glycosidases is reverse hydrolysis. The properties and applications of glycosidases were reviewed some time ago [61]. Transferases are also able to catalyze the transfer of a chemical moiety from a donor to an acceptor. There are many possibilities concerning the combination of glycosyl donor, glycosyl acceptor, and the catalyst enzyme.

Enzymatic synthesis of oligosaccharides by a reverse reaction of hydrolysis is a well-known, generally accepted methodology in the food industry. The position of the equilibrium of a hydrolysis reaction towards synthesis can be shifted by 1) increasing substrate concentration 2) decreasing the amount of water 3) removing the final product from the reaction system (for example, by precipitation or by extraction). As a catalyst for synthesis of oligosaccharides by reverse hydrolysis, several glycosidases have been used [62].

These enzymatic approaches of glycoside synthesis originate from a β-glucosidase-catalyzed condensation reaction of glucose in a concentrated glucose solution [63]. The synthesis of oligosaccharides has been demonstrated by reversing hydrolysis reaction of β-glucosidase at high substrate concentration (90%) [64]. The reversed reaction has also been achieved efficiently by using a column packed with activated carbon for absorption of the resulting product [65]. The condensation reaction between galactose and sucrose was catalyzed by α-galactosidase [66]. Interestingly, rafinose **28** and planteose **29** were formed in a ratio of 3:2 when the reaction was carried out under a batch condition, whereas only rafinose **28** formed using the column. The monosaccharide solution was recycled after the selective adsorption of oligosaccharides onto activated carbon.

The following drawbacks of using glycosidases in reverse reactions of hydrolysis have been pointed out [67]. 1) The yield of the product is normally low. 2) Several isomers which are difficult to separate, are produced. 3) The reaction requires long reaction times and a high enzyme concentration.

5.2 Glycosyl Fluorides as Glycosyl Donors

Glycosyl fluorides, sugar derivatives whose anomeric hydroxyl group is replaced by fluorine atom, are interesting carbohydrate derivatives mainly from their physical organic [68] and biological aspects [69–71]. In 1967, Barnett reported for the first time that a glycosyl fluoride could be recognized by a glycosidase [72]. Since then, numerous studies on the interaction of glycosyl fluorides and enzymes have been reported [73–80]. The use of a glycosyl fluoride as glycosyl donor for an enzymatic glycosylation has the following advantages. First, the fluorine atoms is small enough, like the OH group, to be accepted at an active site of an enzyme, and hence, a glycosyl-enzyme reactive intermediate is easily formed by cleaving the carbon-fluorine bond of the glycosyl fluoride. Second, of the glycosyl halides, only a glycosyl fluoride is stable as an unprotected form, which is necessary for most of enzymatic reactions carried out in the presence of water. In fact, both of α- and β-glycosyl fluorides can exist as considerably stable compounds; other glycosyl halides, glycosyl chloride and glycosyl bromide exist only in the α-form and the opposite β-form is normally thermodynamically unstable due to the "anomeric effect" [81, 82].

The first in vitro synthesis of cellulose via nonbiosynthetic pathway has been achieved by an enzymatic polycondensation of β-D-cellobiolsyl fluoride **30** as substrate for cellulase [83]. Cellulose is a symbolic substance in polymer chemistry because several significant and fundamental concepts in the polymer science field were born through the investigation of cellulose at the early stage of polymer science in the 1920s [84]. In vitro synthesis of cellulose, therefore, has long been a most attractive, challenging dream for synthetic chemists for more than a half century [85–94].

The enzyme catalyst used for the reaction is cellulase, an extracellular hydrolysis enzyme of cellulose. The enzyme promotes transglycosylation of the cellobiosyl moiety toward the 4' hydroxyl group of another cellobiosyl fluoride eliminating hydrogen fluoride. The polymerization was carried out in aqueous

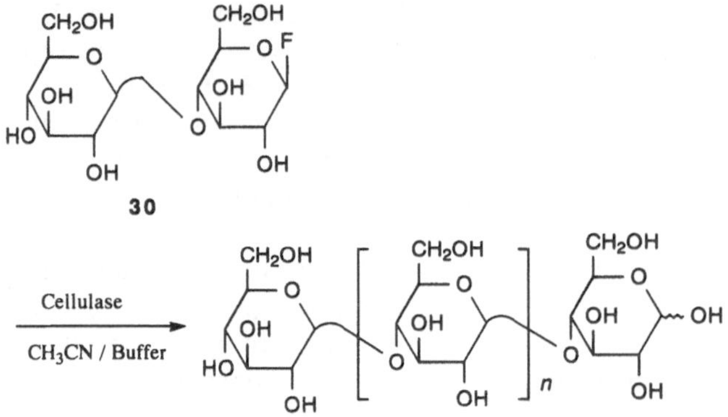

30

organic solvent system in order to make the desired polycondensation predominant in comparison with the competitive hydrolysis reaction. A mixed solvent of acetonitrile/acetate buffer (pH 5) (5 : 1) gave the best result in terms of the yield of water-insoluble "synthetic cellulose". X-ray as well as ^{13}C NMR analyses showed that its crystal structure is of type II with high crystallinity. Under reaction conditions of a higher substrate concentration or higher acetonitrile concentration, water-soluble cellooligosaccharides were produced predominantly.

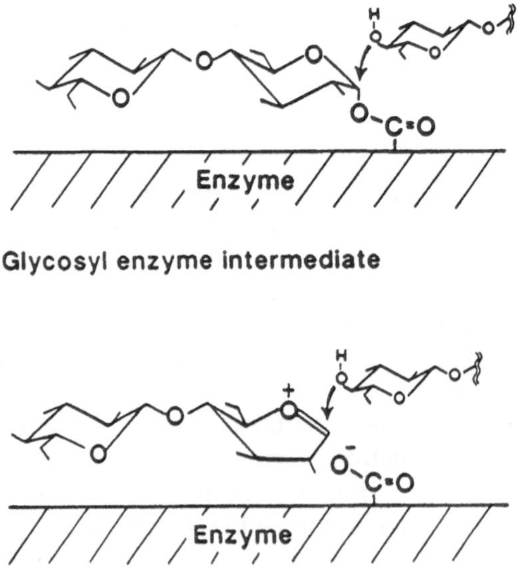

(a) Glycosyl enzyme intermediate

(b) Glycosyl oxocarbenium ion intermediate

Fig. 6. Formation of β(1→4) linkage via substrate enzyme complex involving double inversion of configuration at C1 carbon

The formation of the stereoregular $\beta(1 \rightarrow 4)$ linkage is explained as follows (Fig. 6). The first step involves the formation of a glycosyl-enzyme intermediate or a glycosyl oxocarbenium ion at an active site of cellulase with the elimination of fluoride anion. This reactive intermediate is then attacked by the 4'-hydroxy group of another monomer or propagating polymer which locates in a subsite of the enzyme leading to the stereoselective formation of the $\beta(1 \rightarrow 4)$ linkage. Consequently, the stereochemistry of the product is retention of configuration via "double inversion" concerning the anomeric carbon atom of the β-D-cellobiosyl fluoride.

This reaction mechanism can be compared with a biosynthetic pathway of cellulose that involves the "inversion" of the configuration concerning the C1 carbon atom of the substrate of uridine diphosphate-glucose (UDP-glucose) [95].

In the course of the synthetic study using β-D-cellobiosyl fluoride (Glc-Glc-F) 30, β-D-lactosyl fluoride (Gal-Glc-F), the 4' epimer of 30, was recognized as a substrate for cellulase and is capable of forming a reactive intermediate (Gal-Glc-Enz). By utilizing the transglycosylation activity of the intermediate toward various cellobiose derivatives (Glc-Glc-OR), a new method for selective cellooligosaccharide synthesis has been developed [96]. The reaction consists of two enzymatic procedures; 1) the regio- and stereo-selective lactosylation of cellobiose derivatives as a result of the formation of tetrasaccharide (Gal-Glc-Glc-Glc-OR) and 2) the regioselective cleavage of the terminal glycosidic bond by β-galactosidase affording cellotrisaccharide derivative (Glc-Glc-Glc-OR); one glucose unit elongation can be achieved by repeating this process using the resulting cellooligosaccharide as glycosyl acceptor.

The enzymatic polycondensation of a glycosyl fluoride monomer in aqueous organic solvent media has successfully been applied to the amylase-catalyzed preparation of maltooligosaccharides [97]. Maltooligosaccharides are useful substrates as food additives, medicines, and enzyme substrates for clinical research. Generally, maltooligosaccharides are produced by a degradation reaction of polymers such as amylose, amylopectin, and glycogen [98], whereas only few studies on their production from monomers, e.g. the condensation of a glucose derivative or a maltose derivative, have been reported so far. The polymerization was carried out by stirring a mixture of α-D-maltosyl fluoride 31

31

and α-amylase in an organic solvent-phosphate buffer (pH 6) (2:1) at room temperature. The polycondensation reaction took place effectively when the reaction was carried out in a methanol/buffer (2:1) mixed solvent system.

An unnatural oligosaccharide having an $\alpha(1 \rightarrow 4)$ and $\alpha(1 \rightarrow 6)$ linkage alternately has also been prepared using α-D-maltosyl fluoride as substrate for pullulanase, a hydrolysis enzyme which cleaves the $\alpha(1 \rightarrow 6)$ glycosidic bond in pullulane [99]. The ^{13}C NMR spectrum of the reaction mixture showed the formation of a transglycosylation product of maltotetraosyl fluoride containing an $\alpha(1 \rightarrow 4)$ and $\alpha(1 \rightarrow 6)$ linkage alternately in the main chain. The ^{19}F NMR spectrum of the resulting oligomer indicates that the configuration of the terminal anomeric fluorine atom was of the α-type.

A branched cyclodextrin, 6-O-α-maltosyl cyclodextrin (G$_2$-CD), was produced from α-D-maltosyl fluoride and a cyclodextrin by the transglycosylation reaction catalyzed by pullulanase or isoamylase [100, 101].

Cyclodextrins are well known products obtained by the reaction of starch with cyclodextrin-$\alpha(1 \rightarrow 4)$glucosyltransferase. By use of immobilized cyclodextrin-$\alpha(1 \rightarrow 4)$ glucosyltransferase (silica gel support functionalized with glutardialdehyde), α-glucolsyl fluoride is transformed in high yield, predominantly into cyclodextrin and maltooligomers as side products [102]. The polymerization is carried out in 0.05 M sodium acetate buffer (pH 6.0), and the reaction proceeds under slight shaking at 45 °C with the pH value held constant by neutralizing the liberated hydrogen fluoride with 0.5 M NaOH.

5.3 Glycosides as Glycosyl Donors

Chitinase from *Nocardia orientalis* effectively promoted a transglycosylation reaction on tetra-*N*-acetyl-chitotetraose (GlcNAc)$_4$ and penta-*N*-acetyl-chito-pentaose (GlcNAc)$_5$ to give the corresponding hexa-*N*-acetyl-chitohexaose (GlcNAc)$_6$ and hepta-*N*-acetyl-chitoheptaose (GlcNAc)$_7$, respectively [103]. In this reaction, the substrates of (GlcNAc)$_4$ and (GlcNAc)$_5$ behave both as a glycosyl donor and a glycosyl acceptor. The products of (GlcNAc)$_6$ and (GlcNAc)$_7$ are formed by transferring di-*N*-acetyl-chitobiosyl residues to the acceptor oligomers. Transglycosylation reaction from (GlcNAc)$_5$ to the 4-hydroxyl group of *p*-nitrophenyl 2-acetamido-2-deoxy-β-D-glucopyranoside (PNP-GlcNAc) was achieved by lysozyme catalyst in an aqueous methanol solution (2:3) [104]. Transglycosylation reaction from di-*N*-acetyl-chitobiose to hexa-*N*-acetyl-chitohexaose and hepta-*N*-acetyl-chitoheptaose was efficiently induced through catalysis in the presence of ammonium sulfate at a high concentration. The enzyme also regioselectively synthesized *p*-nitrophenyl 4^5O-β-*N*-acetylglucosaminyl-α-maltopentaoside, which is a useful substrate for assay of human amylase in serum, from di-*N*-acetyl-chitobiose as a donor and *p*-nitrophenyl α-maltopentaoside as an acceptor [105].

5.4 Glucosyl Phosphate as Glycosyl Donor

In addition to the enzymatic polymerizations catalyzed by hydrolases and transglycosidases, a phosphorylase was also found to promote a new enzymatic polymerization. Linear, star-, and comb-shaped polymers having amylase chains of uniform length were prepared using D-glucosyl phosphate as substrate for phosphorylase enzyme [106]. The length of the side chain was controlled by a simultaneous start for all chains using a primer with a minimum length of four glucosyl residues.

6 Enzymatic Oxidation Polymerization Using Peroxidase

A biosynthetic path which involves an oxidation-reduction process plays an important role in maintaining the metabolism of living systems. The enzymes responsible for the oxidation-reduction process are called oxidoreductases which normally require an oxidant or a reductant as reacting chemical species for substrates. Although there have been many publications on enzymatic oxidations or reductions in biochemistry or synthetic organic chemistry, it is only in recent years that oxidoreductases have first been utilized for polymerizations.

6.1 Polymerization of Phenol Derivatives

Peroxidase is an enzyme which catalyzes the oxidation of a donor to an oxidized donor by the action of hydrogen peroxide, liberating two water molecules. For investigations of in vivo reactions of peroxidases, an oxidation reaction with various substrates has been examined. Horseradish peroxidase (HRP) is a single-chain β-type hemoprotein that catalyzes the decomposition of hydrogen peroxide at the expense of aromatic proton donors. HRP has been found to catalyze coupling of a number of phenols and aromatic amines using hydrogen peroxide as an oxidant [107]. In some cases, the reaction produces unidentified polymeric products which are insoluble in water and organic solvents [108,

109]. Other phenol derivatives such as catechol (*o*-diphenol), hydroquinone (*p*-diphenol), or pyrogallol (1,2,3-triphenol) have also been polymerized by peroxidase in phosphate buffer giving polymeric compounds which are insoluble in an aqueous HCl solution [110].

The yield of the polymer from pyrogallol was 20%, however, that from *o*- or *p*-diphenol was less than 5% (Table 1). After further purification of the polymer by dialysis (molecular weight exclusion 12000), the yield of the remaining polymer was very low, less than 1%. Resorcinol (*m*-diphenol) or phloroglucinol (1,3,5-triphenol) did not polymerize under the same reaction conditions.

ESR spectra of the resulting polymers from catechol and *p*-hydroquinone showed signals probably due to semiquinone radicals. In the IR spectra of the polymers, absorbances were observed in the 1200–1000 cm^{-1} region suggesting the presence of different types of ether groups. The presence of carboxylic acid and phenol groups was generally confirmed. ^{13}C NMR spectra of the polymers showed only minor bands in the aliphatic region (< 100 ppm) but very strong peaks in the aromatic region (100–165 ppm). Peaks due to the phenolic carbons were observed around 150 ppm, while those at 170–180 ppm from carboxyl carbons were seen. Detailed analyses of NMR spectra of the polymers indicate: (1) aromatic rings were built into the polymers; (2) the molecular environments of phenolic OH groups changed and (3) some rings were linked via alkyl ether bridges.

Acetaminophen **32** is a widely used analgesic and antipyretic drug. For investigation of in vivo oxidation of **32**, in vitro enzymatic oxidation of **32** using HRP was examined [111, 112]. The reaction was carried out in a phosphate buffer at 25 °C. From the reaction mixture, two dimers (B and E), three trimers (C, G, and F), and one tetramer (D) were identified by ^{1}H NMR spectroscopy (Fig.7). The oligomers were formed primarily through a covalent bond between carbons *ortho* to the hydroxyl group, and to a lesser extent, between the carbon *ortho* to the hydroxyl group and the amino group of another acetaminophen molecule.

A time course experiment of the polymer formation revealed that the polymerization of **32** occurred rapidly after the addition of hydrogen peroxide, and the product formation was almost complete within 0.5 minute from the initiation. Acetaminophen dimer (B) and acetaminophen trimer (c) were the

Table 1. Enzymatic polymerization of monomeric phenols [110]

Monomer	Yield before dialysis			Yield after dialysis	
	(g)	(%)	(g)	(%)	(g)
Catechol	10.8	4.8	0.52	0.63	0.07
Resorcinol	10.0	0	0		
Hydroquinone	16.0	2.3	0.37	0.39	0.06
Pyrogallol	14.6	20.8	3.03	0.33	0.48
Phloroglucinol	10.0	0	0		

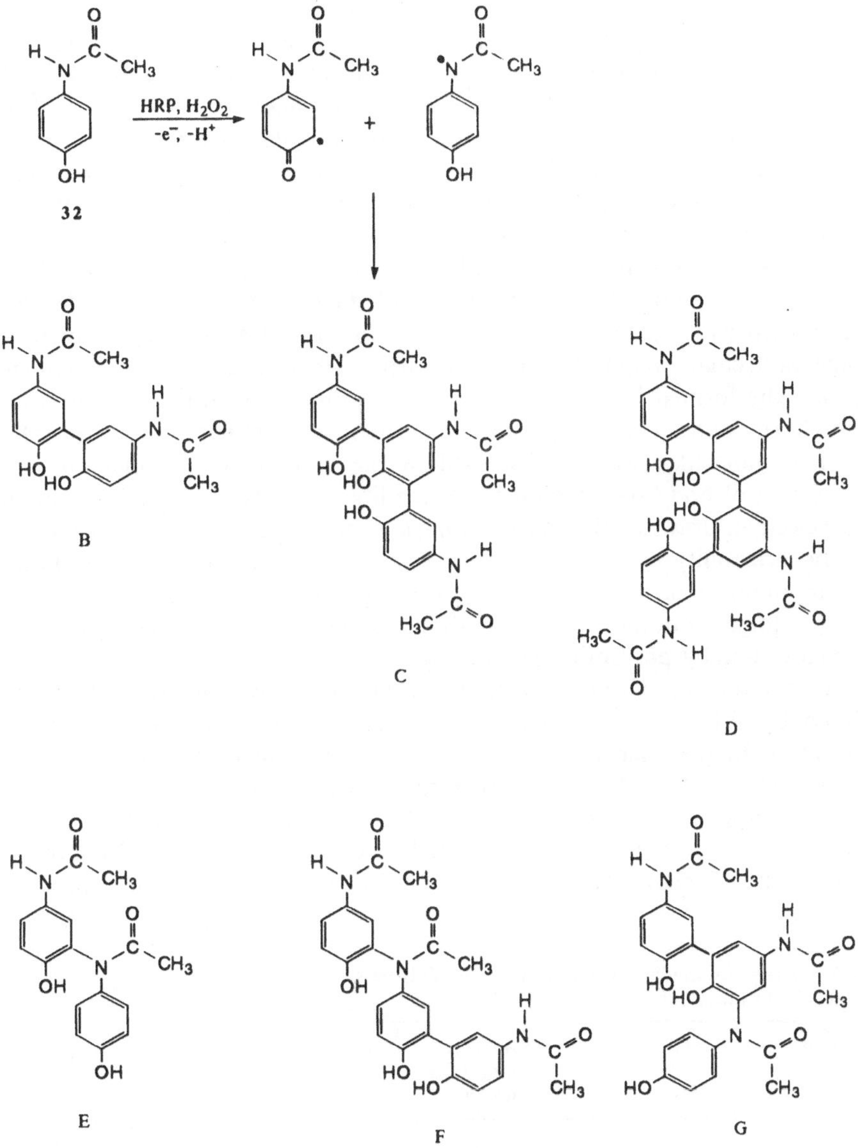

Fig. 7. Structure of products by oligomerization of acetaminophen **32** catalyzed by horseradish peroxidase (HRP)

major products under these conditions. The rapid formation of *N*-acetamino-phen dimer (E) was observed at the initial stage of the reaction and the resulting dimer subsequently decreased, which suggests that it has been further poly-merized giving rise to oligomers of higher molecular weight.

Most of the phenol derivatives are slightly soluble in water and the phenolic oligomers formed are insoluble in water, and hence, enzymatic polymerization in water did not give polymers of higher molecular weight in a good yield. Therefore, the use of organic solvents solubilizing these species is expected to minimize the solubility problems of the enzymatic polymerization.

Polymerization of phenol derivatives using HRP as a catalyst and hydrogen peroxide as an oxidant was explored in a mixture of acetate buffer and a water-miscible organic solvent such as 1,4-dioxane, acetone, *N,N*-dimethylformamide, and methyl formate. Table 2 shows polymerization results of various phenols in 85% dioxane. Among them *p*-phenylphenol **33** gave the highest molecular weight polymer [113]. The majority of the compounds tested yield polymers of average molecular weight in the range from 1400 to 2000. The polymers enzymatically formed from 1-naphthol were insoluble in 1,4-dioxane or *N,N*-dimethylformamide, suggesting a high molecular weight. The reasons for relatively low molecular weights obtained in the case of *p*-chlorophenol, 2,6-dimethylphenol, and 4,4'-biphenol can be readily explained. The first compound has the electron-drawing, ring-deactivating chlorine substituent which decreases the polymerizability by slowing down radical transfer. In the case of the latter two compounds, enzymatic oxidation to the corresponding quinones, 2,2',6,6'-tetramethyldiphenoquinone and 4,4'-diphenoquinone, respectively, probably took place, thereby preventing polymerization.

The structure of the enzymatically synthesized poly(*p*-phenylphenol) was analyzed by FT-IR and CP/MAS ^{13}C NMR spectrometry [114]. The IR spectrum of the polymer indicates that the substitutions on the aromatic ring were at positions 1, 2, 4, and 6 with respect to phenolic-OH. Phenoxy ether linkage was not observed in the spectrum. In CP/MAS ^{13}C NMR spectrum of the polymer, all the aromatic signals were broader than those of the monomer. The major NMR signal at 128 ppm shows that an *ortho*-substituted product

Table 2. Average molecular weights of phenolic polymers enzymatically produced in 85% dioxan [113]

Compound	Average molecular weight
Phenol	1400
p-Methoxyphenol	2000
p-Cresol	1900
p-Chlorophenol	600
2,6-Dimethylphenol	500
4,4'-Biphenol	400
Aniline	1700
1-Naphthol	very high
2-Naphthol	2000
p-tert-Butylphenol	1900
p-Phenylphenol	26000

is the major constituent of poly(p-phenylphenol). These data indicate that the polymer formation in this enzyme catalyzed reactions is mainly through the *ortho* substitutions and to a less extent through *para* substitution on the aromatic ring. This type of condensation reaction could result in a polymer with highly conjugated aromatic backbone with a planar structure. The scope of potential applications of this enzymatic polymerization was significantly expanded by preparation of not only homopolymers but also copolymers. The enzymatic reaction of p-phenylphenol and phenol gave a copolymer containing each monomer unit.

The thermal properties of poly(p-phenylphenol) synthesized by the enzymatic polymerization were examined. TGA of poly(p-phenylphenol) indicated that about 40% of the polymer was lost either by evaporation and/or by degradation on heating the polymer to 600 °C to a stream of nitrogen gas. DSC thermograms of the polymer showed two exothermic heat flows, a minor one (1.14 J/g) at ~ 90 °C and a major one (41.0 J/g) at ~ 200 °C. The melting of the polymer was not sharp, which was also observed with capillary tube determination (the values in the range of 215–300 °C). The two exothermic heat flows observed in the DSC thermogram of poly(p-phenylphenol) could be explained by a crystallization phase followed by branching and/or crosslinking.

The resulting polymer possesses the extensively conjugated π-electrons, and hence, the electrical charge will be carried through the polymer. The surface resistivity of the polymer was found to be ca. 10^5 ω. This value is much higher than that of polyacetylene (10^{-1} ω). Therefore, the phenolic polymer enzymatically produced may be useful as a conductive polymer.

6.2 Polymerization of Aniline Derivatives

Enzymatic polymerization of o-phenylenediamine **34** was performed in a mixture of 1,4-dioxane and phosphate buffer at room temperature [115]. Horseradish peroxidase (HRP) and hydrogen peroxide were used as catalyst and an
oxidizing agent, respectively. The polymerization of **34** in hydrogen peroxide
without the enzyme gave oligomeric products (yield = 67%, Mw = 800). The
polymerization using the enzyme, on the other hand, produced polymers with
higher molecular weight (Mw = 20000) in addition to the oligomers.

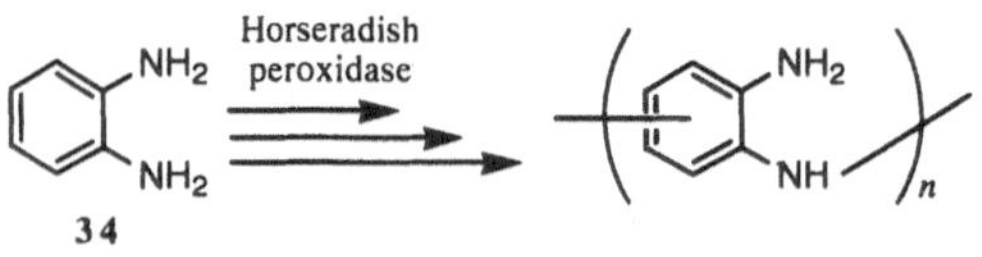

The polymeric portion was separated from the reaction mixture by using
HPLC. From ^{13}C NMR analysis of the polymer, the structure of the polymer
was found to be constructed mainly from imino-2-aminophenylene units:
^{13}C NMR (DMSO-D$_6$) δ 102 (C3 and C6'), 126–128 (C1, C5, C6, C2', C3',
and C4'), 143–145 (C2, C4, C1' and C5') (Fig. 8). In the polymerization of **34** by
a Fe-chelate complex or by electrolysis, the resulting polymer showed mainly a
ladder structure with phenazine rings and contained an insoluble portion. It is
to be noted, however, that the present polymer is of a linear structure and
soluble in a highly polar solvent such as DMF and DMSO.

Enzymatic polymerization of aniline by using another enzyme, bilirubin
oxidase (BOD) has been reported [116]. BOD is a copper-containing oxido-
reductase which oxidizes bilirubin to biliverdin and hydrogen peroxide with
consumption of dissolved oxygen. The enzymatic polymerization took place on
the glass surface of a BOD-adsorbed solid matrix, which was in contact with a
buffer solution containing aniline. The enzymatic reaction markedly depended
upon the concentration of BOD and aniline; under appropriate conditions, a
film-like precipitate formed on the glass surface. The enzymatically synthesized
polyaniline is considered to contain 1,4- and 1,2-substitution structures, which
are partially different form those of chemically or electrochemically prepared
polyaniline.

Scanning electron microscopy (SEM) photographs of the polyaniline film
showed a homogeneous and smooth surface. The cyclic voltammetric studies

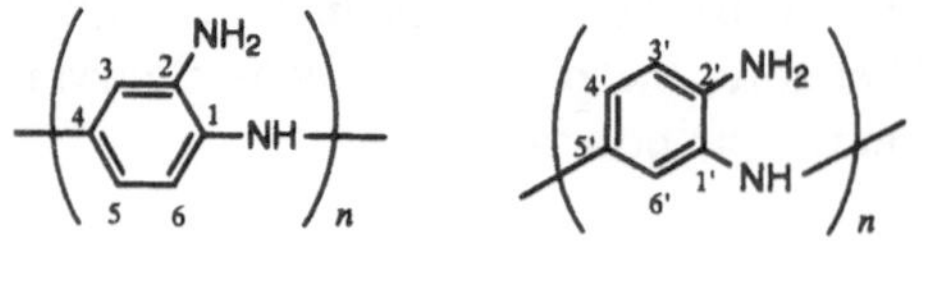

Fig. 8. Structure of polyanilines prepared by
HRP-catalyzed polymerization of o-phenylenediamine

demonstrated that the film was electrochemically reversible in the redox properties in acidic aqueous solution. However, the redox property was different from that of chemically or electrochemically synthesized polyaniline probably due to the structural difference.

BOD was entrapped in the enzymatically synthesized polyaniline film and the enzyme in the film was still active. Therefore, it is expected that this enzymatic synthesis of polyaniline film can be applied in the preparation of electroactive and enzyme-immobilized membranes on various solid matrixes.

7 Conclusion

Even in ancient times, mankind had already started to utilize enzymes for producing cheese, wine, beer, etc. without knowing the existence of the enzyme itself. Even after the development of modern science, it was only at the end of nineteenth century that enzymes were found to catalyze not only hydrolysis in vivo but also its reverse reaction, synthesis. After the significant findings and technical developments in protein chemistry including the elucidation of the mechanism of protein synthesis, the structural determination of protein, the progress of various chromatographies and so on, are we now in a privileged position where many isolated enzymes are available as chemical reagents for various synthetic reactions. In addition, due to genetic engineering including site-directed mutagenesis, it will be possible to produce certain enzymes that are now available only in small quantities. From an industrial viewpoint, further developments will be possible by introducing enzyme immobilization. Much research effort has already been devoted during the last decade to producing satisfactory immobilized enzymes for large-scale applications. Actually, some immobilized enzymes are already in use in industry. Although there have not been many reports concerning "enzymatic polymerization" so far, it is expected that this new concept of polymer synthesis will greatly contribute to the creation of future macromolecular materials which have been difficult to produce by the conventional methodologies.

8 References

1. Bamford CH (1985) In: Mark HF, Bikales NM, Overberger CG, Menger G (eds) Encyclopedia of polymer science and engineering. 2nd edn. Wiley, New York, vol 13, p 708
2. Gandini A, Cheradame H (1985) In: Mark HF, Bikales NM, Overberger CG, Menger G (eds) Encyclopedia of polymer science and engineering. 2nd edn. Wiley, New York, vol 2, p 729
3. Bywater S (1985) In: Mark HF, Bikales NM, Overberger CG, Menger G (eds) Encyclopedia of polymer science and engineering. 2nd edn. Wiley, New York, vol 2, p 1

4. Pino P, Giannini U, Porri L (1985) In: Mark HF, Bikales NM, Overberger CG, Menger G (eds) Encyclopedia of polymer science and engineering. 2nd edn. Wiley, New York, vol 8, p 147
5. Sih CJ, Chen CS (1984) Angew Chem Int Ed Engl 23: 570
6. Luisi PL (1985) Angew Chem Int Ed Engl 24: 439
7. Simon H, Bader J, Gunther H, Newmann S, Thanos J (1985) Angew Chem Int Ed Engl 24: 539
8. Whitesides GM, Wong CH (1985) Angew Chem Int Ed Engl 24: 617
9. Jones JB (1986) Tetrahedron 42: 3351
10. Price NC, Stevens L (1989) In: Fundamentals of enzymology. 2nd edn. Oxford University Press, Oxford
11. Watson JD, Hopkins NH, Roberts JW, Steiz JA, Weiner AM (1987) In: Molecular biology of the gene. 4th edn. The Benjamin/Cummings Publishing Co. Menlo Park, CA, Chap 13
12. Watson JD, Hopkins NH, Roberts JW, Steiz JA, Weiner AM (1987) In: Molecular biology of the gene. 4th edn. Benjamin Cummings, Menlo Park, CA, Chap 14
13. Colvin JR (1959) Nature 183: 1135
14. Bureau TE, Brown RM Jr (1987) Proc Natl Acad Sci USA 84: 6985
15. Elbein AD, Barber GA, Hassid WZ (1964) J Am Chem Soc 86: 309
16. Light DR, Dennis MS, et al. (1989) J Biol Chem. 264: 18589, 18598, 18608, 18618
17. Holmes PA (1985) Phys Technol 16: 32
18. Doi Y, Tamaki A, Kunioka M, Soga K (1988) Appl Microbiol Biotechnol 28: 330
19. Kunioka M, Nakamura Y, Doi Y (1988) Polymer Commun 29: 174
20. Brandl H, Gross RA, Lenz RW, Fuller RC (1988) Appl Environ Microbiol 54: 1977
21. Kirchner G, Scollar MP, Klibanov AM (1985) J Am Chem Soc 107: 7072
22. Klibanov AM (1986) Chemtech 16: 354
23. Laane C, Tramper J, Lilly MD (eds) Biocatalysis in organic media. Elsevier, Amsterdam
24. Halling PJ (1987) Biotechnol Adv 5: 47
25. Zaks A, Russell AJ (1988) Biotechnol 8: 259
26. Deetz JS, Rozzell JD (1988) Trends Biotechnol 6: 15
27. Dordick JS (1989) Enzyme Microb Technol 11: 194
28. Klibanov AM (1989) Trends Biochem Sci 14: 141
29. Klibanov AM (1990) Acc Chem Res 23: 114
30. Matsumura S, Takahashi J (1986) Makromol Chem, Rapid commun 7: 369
31. Gutman AL, Oren D, Boltanski A, Bravdo T (1987) Tetrahedron Lett 28: 5367
32. Okumura S, Iwai M, Tominaga Y (1984) Agric Biol Chem 48: 2805
33. Ajima A, Yoshimoto T, Takahashi K, Tamaura Y, Saito Y, Inada Y (1985) Biotechnol Lett 7: 303
34. Zhi-Wen G, Sih CJ (1988) J Am Chem Soc 110: 1999
35. Margolin AL, Crenne JY, Klibanov AM (1987) Tetrahedron Lett 28: 1607
36. Gutman AL, Bravdo T (1989) J Org Chem 54: 5645
37. Farina M, Modena M, Ghizzoni W (1962) Rend Acad Naz Lincei 32: 91
38. Wallace JS, Morrow CJ (1989) J Polym Sci: Part A: Polym Chem 27: 2553
39. Abramowicz DA, Keese CR (1989) Biotechnol Bioeng 33: 149
40. Riva S, Chopineau J, Kieboom APG, Klibanov AM (1988) J Am Chem Soc 110: 584
41. Patil DR, Rethwisch DG, Dordick JS (1991) Biotechnol Bioeng 37: 639
42. Bodanszky M, Bodanszky A (1984) In: The practice of peptide synthesis. Springer, Berlin Heidelberg New York
43. Erickson BW, Merrifield RB (1976) In: Neurath H, Hill RL, Boeder CL (eds) The protein. Academic, New York vol 3, p 237
44. Merrifield RB (1963) J Am Chem Soc 85: 2149
45. Fruton JS (1982) Adv Enzymol 53: 239
46. Jakubke HD, Kuhl P, Konnecke A (1985) Angew Chem Int Ed Engl 24: 85
47. Kitazume T, Sato T, Kobayashi T (1988) Chem Express 3: 1354
48. Sakuma S, Hayashi T, Ikada Y (1986) Polym Prepr Japan 35: 1850
49. Khan GF, Kobatake E, Shinohara H, Ikariyama Y, Aizawa M (1992) 61st Annu Meet Chem Soc Jpn Prepr II: 1377
50. Ginsburg V, Robbins PW (1984) Biology of carbohydrates, vol 2. Wiley, New York
51. Brady R (1986) Chemistry and Physics of Lipids, vol 42. Elsevier
52. Hakomori S (1981) Annu Rev Biochem 50: 733
53. Hakomori S (1985) Cancer Res 45: 2405
54. Lemieux RU (1978) Chem Soc Rev 7: 423
55. Paulsen H (1984) Chem Soc Rev 13: 15

56. Schmidt RR (1986) Angew Chem 98: 213
57. Kochetkov NK (1987) Tetrahedron 43: 2389
58. Schuerch C (1972) Adv Polym Sci 10: 173
59. Sumitomo H, Okada M (1984) In: Ivin KJ, Saegusa T (eds) Ring-opening polymerization. Elsevier, London, vol 1, p 299
60. Uryu T, Yamanouchi J, Kato T, Higuchi S, Matsuzaki K (1983) J Am Chem Soc 105: 6865
61. Flowers HM, Sharon N (1979) Adv Enzymol 48: 29
62. Monsan P, Paul F, Remaud M, Lopez A (1989) Food Biotechnol 3: 11
63. Bourquelot E, Herissey H, Coivre J (1913) J. Pham. Chim. 8: 441
64. Ajisaka K, Nishida H, Fujimoto H (1987) Biotechnol Lett 9: 243
65. Ajisaka K, Nishida H, Fujimoto H (1987) Biotechnol Lett 9: 387
66. Ajisaka K, Fujimoto H (1989) Carbohydr Res 185: 139
67. Nilsson K (1988) Trends Biotechnol 6: 256
68. Hall LD (1964) Advan Carbohydr Chem 19: 51
69. Bessell EM, Foster AB, Westwood JH (1972) Biochem J 128: 199
70. Barnett JEG (1970) Biochem J 118: 843
71. Walsh C (1983) In: Meisler A (ed) Advan enzymol Wiley, New York, vol V, p 197
72. Barnett JEG, Jarvis WTS, Munday KA (1967) Biochem J 105: 669
73. Card PJ, Hitz WD (1984) J Am Chem Soc 106: 5348
74. Drueckhammer DG, Wong CH (1985) J Org Chem 50: 5912
75. Gold AM, Osber MP (1971) Biochem Biophys Res Commun 42: 469
76. Hehre EJ, Genghof DS, Okada G (1971) Arch Biochem Biophys 142: 382
77. Genghof DS, Brewer CF, Hehre EJ (1978) Carbohydr Res 61: 291
78. Hehre EJ, Brewer CF, Genghof DS (1979) J Biol Chem 254: 5942
79. Okada G, Genghof DS, Hehre EJ (1979) Carbohydr Res 71: 287
80. Hehre EJ, Matsui H, Brewer CF (1990) Carbohydr Res 198: 123
81. Lemieux RU, Koto S (1974) Tetrahedron, 30: 1933
82. Szarek WA, Horton D (1979) (ed) ACS symposium series 87. Washington, DC
83. Kobayashi S, Kashiwa K, Kawasaki T, Shoda S (1991) J Am Chem Soc 113: 3079
84. Mark H (1980) Cellul Chem Technol 14: 569
85. Schlubach HH, Lührs L (1941) Ann 547: 73
86. Husemann E, Müller GJM (1966), Makromol Chem 91: 212
87. Hirano S (1973) Agric Biol Chem 37: 187
88. Schuerch C (1972) Adv Polym Sci 10: 173
89. Micheel F, Brodde O-E, Reinking K (1974) Liebigs Ann Chem 124
90. Micheel F, Brodde O-E (1974) Liebigs Ann Chem 702
91. Uryu T, Kitano, Ito K, Yamanouchi J, Matsuzaki K (1981), Macromolecules 14: 1
92. Uryu T, Yamanouchi J, Kato T, Higuchi S, Matsuzaki K (1983) J Am Chem Soc 105: 6865
93. Uryu T, Yamaguchi C, Morikawa K, Terui K, Kanai K, Matsuzaki K (1985) Macromolecules 18: 599
94. Nakatsubo F, Takano T, Kawada T, Murakami K (1989) In: Kennedy JF, Phillips GO, Williams PA (eds) Celluose, structural and functional aspects. Ellis Horwood, Sussex, p 201
95. Delmer DP (1983) Adv Carbohydr Chem Biochem 41: 105
96. Kobayashi S, Kawasaki T, Obata K, Shoda S (1992) Polym Prepr Japan 41: 1094
97. Kobayashi S, Shimada J, Kashiwa K, Shoda S (1992) Macromolecules 25: 3237
98. Fujita K, Tahara T, Imoto T, Koga T (1988) Chem Lett 1329
99. Kobayashi S, Shimada J, Wen X, Shoda S (unpublished results)
100. Kitahata S, Yoshimura Y, Okada S (1987) Carbohydr Res 159: 303
101. Yoshimura Y, Kitahata S, Okada S (1987) Carbohydr Res 168: 285
102. Treder W, Thiem J, Schlingmann M (1986) Tetrahedron Lett 27: 5605
103. Usui T. Hayashi Y, Nanjo F, Sakai K, Ishido Y (1987) Biochim Biophys Acta 923: 302
104. Usui T, Hayashi Y, Nanjyo F, Ishido Y (1988) Biochim Biophys Acta 953: 179
105. Usui T. Isobe K, Matsui H (1989) Abstracts of XIIth Japanese carbohydrate symposium, 133
106. Ziegast G, Pfannemüller B (1987) Carbohydr Res 160: 185
107. Saunders BC, Holmes-Siedle AG, Stark BP (1964) Peroxidase. Buttersworth, London
108. Josephy PD, Eling TE, Mason RP (1983) Mol Pharmacol. 23: 461
109. Josephy PD, Eling TE, Mason RP (1983) J Biol Chem 258: 5561
110. Schnitzer M, Barr M, Hartenstein R (1984) Soil Biol Biochem 16: 371
111. Potter DW, Miller DW, Hinson JA (1985) J Biol Chem 260: 12174
112. Potter DW, Miller DW, Hinson JA (1986) Mol Pharmacol 29: 155

113. Dordick JS, Marletta MA, Klibanov AM (1987) Biotechnol Bioeng 30: 31
114. Akkara JA, Senecal KJ, Kaplan DL (1991) J Polym Sci, Polym Chem Ed 29: 1561
115. Kobayashi S, Kaneko I, Uyama H, (1992) Chem Lett 393
116. Aizawa M, Wang L, Shinohara H, Ikariyama Y (1990) J Biotechnol 14: 301

Editor: Prof. Ringsdorf
Received October 1992

Soluble Polymer Supports for Liquid-Phase Synthesis

K.E. Geckeler
Institute of Organic Chemistry, University of Tübingen,
Auf der Morgenstelle 18,
D-72076 Tübingen, Germany

This review presents a survey on functional soluble polymers in view of their use as supports for liquid-phase synthesis. The general aspects of synthesis in homogeneous media as well as analytical and separation problems are discussed, focussing on the role of the polymer in the synthetic cycle and the problems associated with polymer-supported reactions. A survey of polymeric carriers in respect of their functional groups and backbones is provided with an emphasis on poly(oxyethylene), polystyrene, and poly(vinyl alcohol) supports. Combined methods using solid and soluble supports are also highlighted. The polymeric carriers are discussed and evaluated for their use in peptide and nucleotide synthesis. Finally an outlook into future developments is attempted.

Advances in Polymer Science, Vol. 121
© Springer-Verlag Berlin Heidelberg 1995

List of Symbols and Abbreviations

A	Amino acid or Adenosine
Ac	Acetyl
AIBN	2,2'-Azobisisobutyronitrile
Ala	Alanine
Arg	Arginine
Asn	Asparagine
Boc	*tert*-Butyloxycarbonyl
bz	Benzoyl
Bzl	Benzyl
Cys	Cysteine
d	2'-Deoxyribonucleoside
DCC	*N,N'*-Dicyclohexylcarbodiimide
DCM	Dichloromethane
DMF	*N,N'*-Dimethylformamide
dT	Deoxythymidine
Glu	Glutamic acid
Gly	Glycine
HAc	Acetic acid
Ile	Isoleucine
Leu	Leucine
$\bar{M}$	Average molecular mass
Met	Methionine
NMM	*N*-Methylmorpholine
Np	Nitrophenyl
Nuc	Nucleotide
P	Polymer
p	Phosphoric acid residue
pdT	2'-Deoxythymidine monophosphate
PEG	Poly(ethylene glycol); α,ω-Dihydroxy poly(oxyethylene)
Phe	Phenylalanine
POE	Poly(oxyethylene); mono- or bifunctional
Pro	Proline
PS	Polystyrene
PSC	Polystyrene, crosslinked
PVA	Poly(vinyl alcohol)
Ser	Serine
T	Thymidine
Th	Thymine

THF	Tetrahydrofuran
TPS	Triisopropyl benzenesulfonyl chloride
Tos	Tosyl; 4-Methylbenzenesulfonyl
Trit	Trityl
U	Uracil
Val	Valine
X	Functional group
Y	Protecting group
Z	Benzyloxycarbonyl

1. Introduction

Nowadays there are three principal routes in repetitive-type synthesis: the classical method, which involves laborious purification steps, the use of cross-linked matrix polymers or the application of soluble polymer supports. Both polymer-based methods impress by their ease of handling and their versatility and flexibility in the general strategy of stepwise synthesis.

Being based on solid-phase synthesis in principle, the liquid-phase method for the synthesis of biopolymers is characterized by the same limitations due to polymer reactions. These mainly concern the attachment and release from the polymer which are the most crucial steps in polymer-supported synthesis. However, additional problems are caused by the heterogeneity of the reactions which decreases the yield considerably. The main reason why this concept has found a widespread application is the ease of experimental procedure, using filtration for separation, thus avoiding the laborious and time-consuming procedure of separation of products from reagents during repetitive synthesis [1–7].

Many approaches in liquid-phase synthesis have been made to mimic the characteristics of crosslinked polystyrene which is used in solid-phase synthesis as a basis support. The excellent properties of this quasi-ideal carrier introduced by Merrifield in 1963 have not been attained in a comparable manner for liquid-phase synthesis so far [1, 5]. These properties are:

– easy functionalization and variation of functional capacity over a broad range,
– good mechanical and chemical stability, and
– commercial availability or rapid and convenient preparation.

The usual matrix consists of polystyrene which is crosslinked by 1 or 2% of divinylbenzene. The degree of crosslinking regulates the swellability of the resin which has a considerable influence on solvation of the polymer and its side-chains. The physicochemical incompatibility of polystyrene supports is understandable when regarding the polymer-supported peptide as a grafted hydrophilic side chain, which increases during stepwise synthesis, on a hydrophobic backbone. Therefore, attempts have been made to increase hydrophilicity of the crosslinked polymer matrix [8, 9]. Thus, by approaching the structure-based physicochemical properties of the polymer backbone a similar solvation effect could be eventually obtained in order to provide maximum yields of the heterogeneous polymer-analogous reactions.

Encouraged by the striking advantages of the use of insoluble polymer carriers the application of soluble polymer supports has found increasing interest in many different fields [10–20]. Of particular mention in this context – in addition to the synthesis of peptides and oligonucleotides – seems to be the use of soluble polymers in the synthesis of oligosaccharides, their use as organic and complex forming reagents and for catalysis [10, 16–20].

2 General Aspects of the Application of Soluble Polymer Supports

2.1 Requirements and Selection of Basis Polymers

The first important step in polymer-supported repetitive-type synthesis is the attachment of the first synthetic unit to the carrier. In principle, two types of functional supports have to be considered:

   P–X

   P–X–A–Y

X = functional group
A = anchor group
Y = terminal group

If there is a functional basis polymer available which meets the requirements of attachment in terms of the functional group, the polymer support can be used without further modification. The introduction of an anchor group permits, in addition, the alteration of the polymer-peptide bond. This provides a greater flexibility for the selection of side-chain protecting groups. As there is a special section dedicated to this topic, no details are given here.

The use of soluble polymers in repetitive-type synthesis is based on several prerequisites:

- Easy preparation or commercial availability
- Adequate functional capacity
- All educts and products must be soluble to permit homogeneous reactions
- Kinetics should be favorable for all type of reactants, mono-, bi-, or multifunctional compounds.
- Steric factors and diffusion should be sufficiently taken into account—this leads to the optimization of concentration and temperature of the polymer solution.

A basic requirement of a polymer support so that it is compatible with its use in liquid-phase synthesis of the repetitive sequential-type is solubility in a variety of solvents and preferably in water [10]. Also, when not using mono- or bifunctional supports, in the case of multifunctional polymers, an alternating arrangement of the functional sites along the macromolecule is advantageous. Otherwise, sterical problems caused by neighbouring effects could considerably decrease the yields of the polymer-analogous reactions.

The classical investigations by Flory into the kinetics of esterification using polymers showed the independence of the size of macromolecules in terms of the reactivity of the polymer-attached functional group [21]. Interestingly, comparable studies carried out with soluble polymers were in accordance with his findings [22, 23].

Differences between linear, soluble polymers and cross-linked matrix supports in terms of their kinetic aspects were first reported by Andreatta and Rink [24]. In comparison with the corresponding monomeric analogous compound, they found substantial differences for an aminolysis reaction which can be explained by interactions with the polymer and steric hindrance of the reactive sites (Table 1).

A wide range of soluble polymers are suitable in principle for their application as a polymer support: poly(ethylene glycol), poly(vinyl alcohol) and copolymers, poly(ethylene imine), polyacrylamide, polystyrene, and their copolymers. During the last decade quite a number of studies have been carried out in order to find new types of carriers or at least to improve the properties of available polymer supports (Table 2). However, the unique features of soluble supports based on poly(oxyethylene) could not be reproduced by another polymer type. Some progress has been made by several refinements and by introduction of anchor groups. It can be stated in general that, up to now, poly(oxyethylene)-based derivatives have proved themselves to be the most appropriate soluble polymer supports for use in repetitive-type synthesis, although in a few cases specially designed polymers can be more advantageous.

2.2 Salient Features of Poly(oxyethylene) Carriers

Poly(oxyethylene) derivatives frequently used are bifunctional, i.e.,

$$X-(CH_2CH_2O)_n CH_2CH_2-X$$

For brevity and due to the fact that in many cases maximum functional capacity (100% = 2 mol/molecule) is not achieved, in this chapter the following abbreviated formula is also used:

POE–X

In many cases commercially available poly(ethylene glycol) is used without further purification by special methods prior to use. The molecular mass ($2000 < \bar{M}/(\text{g mol}^{-1}) < 20\,000$) was selected according to the loading capacity desired and the hydrophobicity of the peptide sequence to be synthesized.

One of the salient features of PEG is its bifunctionality which involves that the growing product chains are not sterically hampered by neighbouring

Table 1. Second order rate constants for the aminolysis of Z–Ala–ONp with H–Pro–O–CH$_2$–R in chloroform at 24 °C [24]

R	K_2 $(\text{L mol}^{-1}\text{s}^{-1})$
Phenyl	0.081
Polystyrene, linear $\bar{M} = 20\,400$	0.047
Polystyrene, crosslinked with 2% divinylbenzene	0.030

Table 2. Synthetic basis polymers for liquid-phase repetitive-type synthesis

Polymer backbone	Functional group	Type of synthetic product	Ref.
(CH_2-CH_2-O)	$-OH$	Peptide	25–27
	$-NH_2$	Oligonucleotide,	28, 29
		Peptide	30
	$-Trit-Cl$	Oligonucleotide	31
	$-CH_2COOH$	Peptide	32, 33
$(CH-CH_2)$ $\mid$ OH	$-OH$	Oligonucleotide	34
$(CHCH_2)(CHCH_2)$ N—O OH	$-OH$	Peptide Oligonucleotide	15, 35 36
(CH_2-CH_2-NH)	$-NH_2$	Peptide	37, 38
$(CHCH_2)$ ⬡	$-CH_2Cl$ $-Trit-Cl$	Peptide Oligonucleotide	24, 39 40, 41

molecules. The relationship between functional capacity and molecular mass of PEG is outlined in Table 3.

Another outstanding property of POE-based polymers is the solubility in both water and many organic solvents. Table 4 allows us to assess the solubility of POE polymers in some solvents. Although a series of solvents can be used for PEG supports, very frequently dichloromethane, dimethylformamide, and pyridine were used for the attachment of the first synthetic unit or for repetitive coupling during synthesis.

2.3 Separation of Reagents from the Polymer Support

A major problem which has to be faced in all repetitive-type syntheses where excess reagents are used is the purification of the polymer or the removal of low-molecular compounds. Beside precipitation of the support, which was applied even in the beginning of the use of soluble polymers before being refined later by the introduction of crystalline polymers, in many studies on liquid-phase application, membrane filtration was also used for the separation of low- and high-molecular components in solution [10, 25–28, 42–48]. The experimental arrangement of a typical membrane filtration system is depicted in Fig. 1.

Even though precipitation is nowadays used in general, after some synthetic cycles a purification step by ultrafiltration can be recommended in order to

Table 3. Dependence of functional capacity and melting point on molecular mass of poly(ethylene glycol)

$\dfrac{\bar{M}}{10^3 \text{ g mol}^{-1}}$	m p. in °C	Hydroxyl number in mg KOH g^{-1}	Nominal functional capacity in mmol g^{-1}
1	35–40	107–118	2
1.5	44–48	70–80	1.33
2	48–52	51–62	1.0
3	51–55	34–42	0.66
4	53–58	23–30	0.50
6	55–60	16–20	0.33
10	55–60	9–13	0.20
20	ca. 60	< 7	0.10

Table 4. Solubility of poly(ethylene glycol) ($\bar{M} = 4000$ g mol^{-1}) in some solvents [42]

Solvent	Solubility in wt.-%
Water	55
Dichloromethane	53
Ethanol (60%)	50
Chloroform	47
Pyridine	40
Methanol	20
Benzene	10
1,4-Dioxane	10
Carbontetrachloride	10

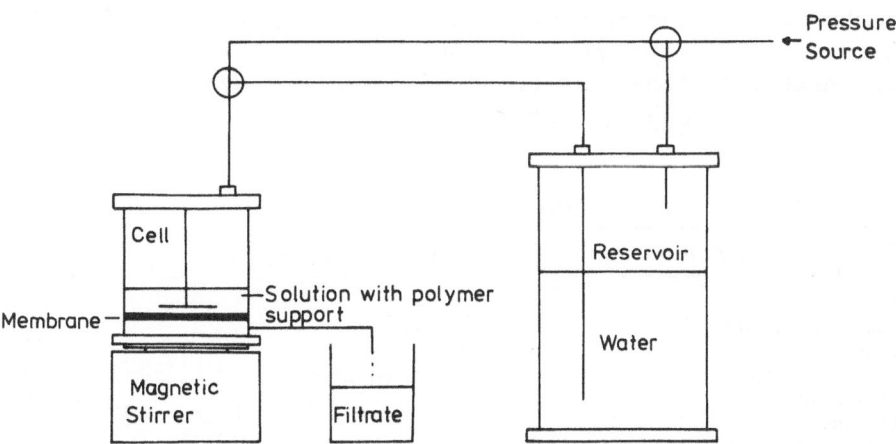

Fig. 1. Membrane filtration system for the separation of excess reagents from the soluble polymer support during repetitive-type synthesis

provide effective and quantitative separation. Whereas in the beginning of the development of the liquid-phase method, ultrafiltration and diafiltration were preferred for the purification and removal of the polymer support [27], further advancements led to even more advantageous procedures due to the excellent properties of poly(ethylene glycol). One of these particular properties of PEG is its crystallization tendency and the crystalline structure of PEG-peptides which reduce the danger of inclusions during precipitation procedures. Thus, crystallization of the polymer was introduced to the repetitive steps of the synthetic cycles for the removal of excess reagents and allowed the reduction of the period of one synthetic step to a maximum of two hours [49]. Some solvents for recrystallization and precipitation are listed in Table 5.

In this context, the extreme temperature gradient of the solubility of PEG is illustrated in Fig. 2. No essential effect of the molecular mass of PEG on the solubility of polymer-supported peptides was observed in the range between 2000 and 20 000 $g\,mol^{-1}$. Limitations are given by the low functional capacities of the high-molecular weight PEGs and the unfavorable properties of low-molecular glycol supports with $\bar{M} < 2000\ g\,mol^{-1}$ in terms of crystallinity and melting points.

2.4 Analysis of Soluble Supports

In contrast to the various restrictions of analysis in solid-phase synthesis, the analytical control of polymer-supported soluble molecules permits the application of most of the usual analytical methods. This is one of the salient advantages of soluble carrier-supported synthesis. Principally, two main groups of methods for functional group analysis can be distinguished: Spectrometric and chemical methods (Tables 6, 7).

Beside spectrometric analysis, functional groups of polymers which are reactive to form a covalent bond easily can be determined by means of chemical reactions with the appropriate reagents. A prerequisite is the possibility of analysis of the products thus obtained. That means in many cases the introduction of a spectroscopically active function.

Table 5. Solvents for poly(ethylene glycol)

Precipitation by diethylether from	Recrystallization from
Dichloromethane	Ethanol
Dimethylformamide	Methanol
Dimethyl sulfoxide	Ethyl acetate
Methanol, Ethanol	Methylethyl ketone
Tetrahydrofuran	
Pyridine	
Acetic acid	
Trifluoroacetic acid	
1,4-Dioxane	

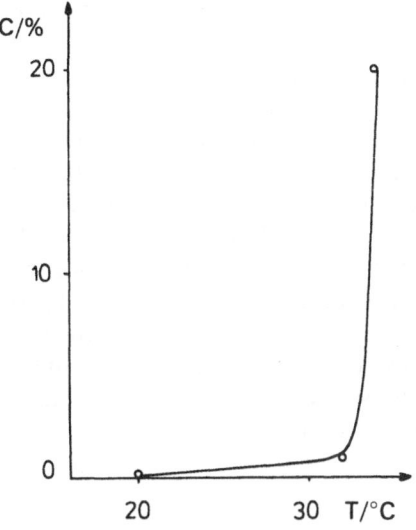

Fig. 2. Solubility dependency of poly(ethylene glycol) ($\bar{M} = 20\,000\,g\,mol^{-1}$) on temperature in ethanol (c = concentration in weight-%)

Table 6. Analytical methods for determination of functional groups on soluble polymer supports

Functional group	Method of determination	Ref.
Hydroxyl	UV spectrometry (after reaction with reagent)	51
Amino	Titration	28
	$^{13}C\,NMR$ spectrometry	52
	IR/UV spectrometry	53, 54
	Amino acid analysis	55, 56
Carboxyl	Amino acid analysis (after coupling and cleavage)	33, 56
Chloromethyl	Elemental analysis	24, 57

Table 7. Chemically-based analytical methods for soluble polymer supports

Analytical method	Functional group	Ref.
Titration	–COOH	58
	–NH$_2$	28, 55
Urethane formation	–OH	51
Amino acid analysis	–COOH	33, 56
	–NH$_2$	55, 56
Paper chromatography	TpT	48, 59
		60

An interesting approach for the analysis of the functionalization of soluble polymer supports could be the introduction of radiolabeled reagents. For a detailed description of the application of these methods see the respective chapters in [50].

In many cases the functional or loading capacity is expressed in $mmol\,g^{-1}$ polymer. For polymer-analogous derivatization reactions the definition of the degree of functionalization (DF) is useful. It is defined as the quotient of the content of new functional groups and that of original functional groups [61].

A general problem of polymer analysis is the distribution of molecular mass which should be as uniform as possible. A certain limitation is given by the purification methods: Ultrafiltration is carried out with membranes which are characterized by a certain molecular mass exclusion. That means no significant change of distribution characteristics of the support during synthesis can be assumed although this has not been confirmed by published studies yet.

3 Polymer Supports for Liquid-Phase Peptide Synthesis

3.1 General Aspects

3.1.1 Role of the Support in the Synthesis Cycle

Peptide synthesis in solution according to the classical method involves three steps, namely the coupling of the first two amino acids, cleavage of the protecting group of the amino function, and the next coupling step. If polymer supports are used, two additional steps are required: one is the attachment of the first amino acid of the peptide sequence to the polymer and the other is the cleavage of the peptide from the polymer support.

The principle of the application of soluble polymer supports in peptide synthesis is the following: The carboxyl group of the C-terminal amino acid of the peptide is attached to a soluble polymer support, thus acting as a macromolecular, solubilizing carboxyl protecting group (Scheme 1).

This soluble polymeric protecting group keeps the peptide attached in solution, even in the case of longer or hydrophobic peptides. Therefore, all coupling steps can be carried out in the homogeneous phase. After the attachment of the first amino acid to the carrier, the unreacted functional groups should be blocked in order to avoid further reactions yielding shorter sequences of the peptide as in solid-phase synthesis [62, 63]. For the frequently used hydroxyl groups, acetic acid anhydride or an isocyanate are convenient blocking reagents. However, as the coupling conditions for the formation of the peptide linkage are milder than those for the attachment to the polymer, this step is often omitted. The influence of amino acid residues on the yield of attachment reaction to the polymer support can be assessed from the data of some examples given in Table 8.

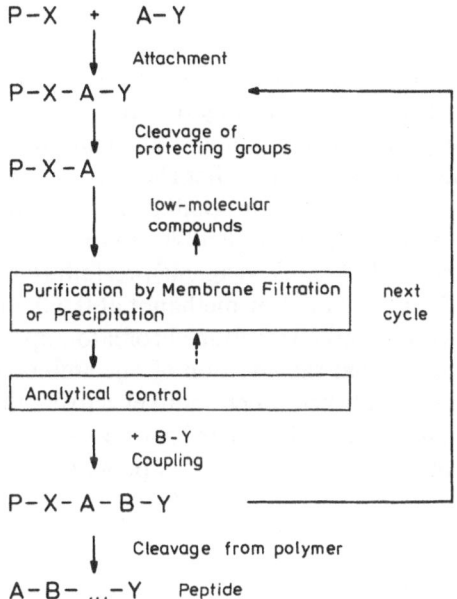

P = Polymer
X = functional groups
A–Y, B–Y = N-protected amino acids

Scheme 1. Scheme of peptide synthesis using soluble polymer supports

Table 8. Attachment of some amino acids to poly(ethylene glycol) ($\bar{M} = 10\,000\ \text{g mol}^{-1}$) using DCC as coupling reagent [27]

Amino acid (Boc or Z)	Number of equivalents	Temperature in °C	Reaction time in d	Yield in %
Gly, Ala	5	20	2	90
Leu, Ile, Val, Met, Phe	10	40	5	80
Pro, Asn (Y)	10	40	7	60

Y = 4,4′-Dimethoxybenzhydryl

Then, the amino protecting group is cleaved and the product purified by membrane filtration or precipitation. After analytical control the next N-protected amino acid is coupled to the amino group of the first unit, or, if the purity of the product is not satisfactory, the purification step is repeated. This synthetic cycle is reiterated until the complete amino acid sequence is obtained. Finally, the N-protected or free peptide can be isolated after one or two cleavage steps.

The final cleavage of the peptide synthesized from the polymer support can be accomplished by alkaline hydrolysis, transesterification, aminolysis, or hydrazinolysis. After cleavage of the peptide from the polymer, the water-insoluble protected peptide is separated by filtration or centrifugation from the polymer support, which has been previously dissolved in water. When the peptide is soluble in ethanol, the product is dissolved in ethanol after cleavage and then the polymer support is removed by crystallization during cooling of the solution. As solvents, dichloromethane, dimethylformamide, dimethyl sulfoxide or their mixtures are recommended. For removal of excess reagents precipitation (30% w/v ether), recrystallization (5–10% w/v ethanol or methanol at 0 °C), ultrafiltration, gel permeation chromatography, and adsorption chromatography have been suggested. It has been pointed out that the addition of one amino acid to the growing peptide chain can be accomplished much more effectively than by classical procedures although new and improved carriers have become available [64–66]. In addition, monitoring of the polymer-supported peptide is easier in the homogeneous phase.

3.1.2 Problems of Polymer-Supported Reactions

Major problems and restrictions in liquid-phase peptide synthesis based on polymer supports are as follows:

- Inadequate structure of the polymer backbone or sidechains: The hydrophilic-hydrophobic balance should be adapted to that of the peptide chain to be synthesized. In addition, the solubilizing power of the macromolecule should be sufficient to solubilize the oligo- or poly-peptide in all solvents and under all conditions envisaged during synthesis.
- Attachment of the first amino acid to the polymer: Incomplete polymer – analogous reaction – in most cases esterification – is not as important as it was considered at the beginning of the liquid-phase synthesis, but is still as essential as it is for matrix reactions [67]. The unreacted functional groups can be blocked to prevent the formation of unwanted peptides. However, for particular amino acids (e.g. proline) the fixation to the polymer still represents a crucial step.
- Stability of the polymer-peptide bond: It has always been a general problem in polymer-supported repetitive-type synthesis to find an adequate linkage which allows convenient final cleavage of the peptide and is simultaneously not affected during cleavage of the amino protecting groups after each synthetic step [68].

Studies by Bayer et al. [22] and Mutter [69] have shown that liquid-phase synthesis of peptides exhibits kinetic behaviour analogous to classic peptide synthesis. The reaction rates of both methods are of the same order of magnitude, and using poly(ethylene glycol) esters, the polymer reaction proceeds at an even higher rate than the reaction involving the corresponding low-molecular compounds. It is also astonishing that there is no dependence of the rate constants on the chain length of the PEG residue (45 < n̄ < 450). Second order kinetics have been found for aminolysis of glycine esters (Table 9).

Generally, research efforts are still needed for polymer preparation and modification to reduce these problems of liquid-phase synthesis to negligible levels.

3.1.3 Survey of Polymeric Carriers

A brief survey of soluble polymer supports classified by the type of polymer backbone is given in Table 10. Some of the soluble supports have already been investigated for automatization of liquid-phase synthesis, e.g. poly(ethylene glycol)s of $\bar{M} \doteq 4000$ and $10\,000$ g mol^{-1} [70].

Table 9. Rate constants of the aminolysis of Boc–Gly–ONp with H–Gly–OR in acetonitrile at 25 °C [22]

R ($\bar{M}$ in g mol^{-1})	K_2
	L mol^{-1} s^{-1}
Ethyl (29)	0.013
tert-Butyl (57)	0.024
Methoxyethyl (59)	0.008
POE (2 000)	0.018
POE (4 000)	0.019
POE (6 000)	0.016
POE (10 000)	0.014
POE (20 000)	0.014

Table 10. Soluble polymer supports in liquid-phase peptide synthesis

Polymer (backbone)	Functional group	$\bar{M}$	Ref.
		(10^3 g mol^{-1})	
POE	–OH	2–20	25–27, 45, 49
	–COOH	6–7	33
	–NH$_2$	3	71, 72
		5	53
		6	16
Poly(1-vinyl-2-Pyrrolidinone-co-vinyl alcohol)	–OH	33	15
Poly(1-vinyl-2-Pyrrolidinone-co-vinylamino acid ester)	–NH$_2$	98	73
Poly(ethylene-imine)	–NH$_2$	30–40	37, 38
Polystyrene	–CH$_2$Cl	200	24, 39

3.2 Poly(oxyethylene)-Based Supports

3.2.1 Survey of Telechelic Carriers

The salient features of POE-derivatized supports are outlined in a previous section. Here, a brief survey of such carriers is listed in terms of the terminal functional groups (Table 11).

3.2.2 Amino and Carboxy Derivatives

A convenient pathway of synthesis of POE-NH$_2$ was offered by Mutter [16]. He prepared the tosylate of PEG and used phthalimide as a masked amino group with subsequent release by hydrazinolysis:

$$\text{POE–OH} \xrightarrow{\text{Tos–Cl}} \text{POE–OTos} \xrightarrow[\text{2) H}_2\text{N–NH}_2]{\text{1) K–Phthalimide}} \text{POE–NH}_2$$

Noteworthy for this two-step derivatization procedure is the high degree of functionalization of ca. 90% as reported for the POE–NH$_2$ derivatives of $\overline{M}$ = 3000 and 4000 g mol^{-1}. This aminofunctionalized POE, prepared by identical or similar routes of synthesis, has also found other applications [83–85]. In addition, suggestions were made for POE-bound active esters of the following type for peptide synthesis [16, 30]:

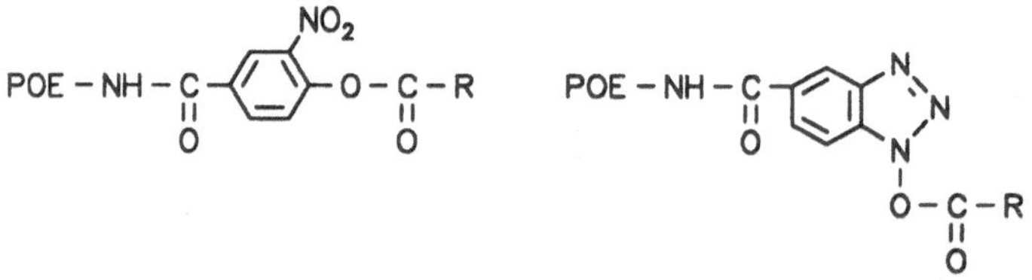

Very few attempts have been made to use soluble polymer supports in the synthesis of peptides for extension of the carboxyl terminus. One example has been published by Royer and Anantharamaiah [33]. They used carboxymethyl poly(oxyethylene) for the synthesis of peptides in aqueous phase with water-soluble coupling agents and immobilized carboxy-peptidase Y for cleavage of the protecting groups:

$$\text{POE–OH} \xrightarrow[\text{1 N NaOH}]{\text{BrCH}_2\text{–COOC}_2\text{H}_5} \text{POE–OCH}_2\text{–COOH}$$

$$\xrightarrow[\text{2) 0.1 N NaOH}]{\text{1) H}_2\text{N–CH}_2\text{–COOCH}_3} \text{POE–OCH}_2\text{–CONH–CH}_2\text{–COOH}$$

The loading capacity was found to be 0.1 mmol g^{-1} as determined by amino acid analysis of glycine. Cleavage of the peptides from the polymer support was performed with cyanogen bromide at room temperature and gave, for example,

Table 11. Poly(ethylene glycol) derivatives (POE-R) for liquid-phase peptide synthesis

R	$\bar{M}$ (10^3 g mol^{-1})	Polymer-bound amino acid	Functional capacity	Ref.
-OH	3	Gly	75%	71, 72
	4	Val		74
	6	Val		49
	10	Gly, Ala, Leu, Ile, Val, Met, Phe, Pro	60–90%	27
	15	Gly, Phe, Met, Val	80%	25, 45, 49
	20	Gly		13
-OCH$_2$C(CH$_3$)$_2$-OH	2	Gly, Thr		75, 76
-NH$_2$	4	Gly, Ala, Val	81–100%	16
	5	Ala, Gly		77
	5	Ala, Tyr	99; 95.1%	53
-CH$_2$COOH	6–7	Gly	0.1 mmol g^{-1}	33
-OCO-C$_6$H$_4$-CH$_2$Br	6	Ala	98%	78
		Gly	105%	
		Phe	92%	
-OCO-C$_6$H$_3$(NO$_2$)-CH$_2$Br	6	Val	0.117 mmol g^{-1}	79, 80
-OCONH (CH$_2$)$_4$O-C$_6$H$_3$(NO$_2$)-CH$_2$Br	6	Gly	0.051 mmol g^{-1}	78
-OCO-CH$_2$-C$_6$H$_4$-CO-CH(CH$_3$)-Cl	6	Tyr	0.18 mmol g^{-1}	81
-OCONH (CH$_2$)$_6$Cl	6	Gly	0.137 mmol g^{-1}	78
-NHCO-C$_6$H$_4$(NO$_2$)-CH$_2$-NH$_2$	5	Leu	0.127 mmol g^{-1}	82

the tetrapeptide Ala-Ala-Cys-Lys in 42% yield. Thus, this interesting new approach proved to be applicable in the synthesis of oligopeptides containing amino acids with side-chain protection.

It is evident, that, because of racemization effects, no comparable approach according to the Letsinger method in solid-phase synthesis, i.e. synthesis from the amino terminus of the peptide has been reported [2]. This would require polymer supports with different functional groups than those using the conventional carboxyl end of the peptide.

Derivatives of PEG, which could be of interest for Letsinger-analogous liquid-phase peptide synthesis, were suggested by Geckeler and Bayer using the following route of synthesis for carboxy polymers [58, 86–88]:

$$POE-OH \xrightarrow{Na} POE-ONa \xrightarrow{Br-CH_2CH_2-COOR}$$

$$POE-O-CH_2CH_2-COOR \xrightarrow{OH^-} POE-O-CH_2CH_2-COOH$$

$$POE-OH \longrightarrow POE-O-\overset{O}{\underset{\|}{C}}-CH_2CH_2-COOH$$

Schoknecht et al. [71, 72]. synthesized the sequence 13–20 of the B chain of insulin using the amino derivative of POE with an anchoring group. This derivative was prepared by the reaction of PEG ($\bar{M} = 3000\,g\,mol^{-1}$) with N-protected glycine and dicyclohexyl carbodiimide as coupling reagent and subsequent cleavage of the protecting group:

$$POE-OH \xrightarrow[DCC]{HOOC-CH_2-NH-Boc} POE-OCO-CH_2-NH-Boc$$

$$\xrightarrow{1)\ 2\ N\ HCl/HAc} POE-OCO-CH_2-NH_2$$

3.2.3 Miscellaneous Supports

Block copolymers as solubilizing protecting groups in peptide synthesis were prepared by Bayer et al. [51]. They used 3,5-diisocyanatobenzyl chloride and 3-nitro-3-azapentane-1,5-diisocyanate as reagents for the synthesis of the copolymeric supports:

$$\text{OCN} - (CH_2)_2 - \underset{\underset{NO_2}{|}}{N} - (CH_2)_2 - NCO \xrightarrow{HO-POE-OH}$$

$$+ O - POE - O - \underset{\underset{O}{||}}{C} - NH - (CH_2)_2 - \underset{\underset{NO_2}{|}}{N} - (CH_2)_2 - NH - \underset{\underset{O}{||}}{C} + \xrightarrow{H_2, Pd/C}$$

$$+ O - POE - O - \underset{\underset{O}{||}}{C} - NH - (CH_2)_2 - \underset{\underset{NH_2}{|}}{N} - (CH_2)_2 - NH - \underset{\underset{O}{||}}{C} +$$

The average molecular mass of the POE blocks in the study mentioned were 400, 1000, 6000 and 10000 g mol^{-1}. Molecular masses of the copolymers prepared by polyaddition were reported to be in the range of 50000–1000000 g mol^{-1}.

A prerequisite for a tailor-made preparation of such a type of block copolymer is an exact analytical determination of the hydroxyl number of PEG. To this end, a method was described which is based on the reaction of PEG with phenylisocyanate and subsequent UV-spectrometric analysis of the urethane formed. The benefit of this type of polymeric support results from the high loading capacity of the block copolymer which cannot be obtained with PEG. In addition, the defined distances of the functional groups on the polymer chain should be advantageous namely for longer peptides. Thus, steric problems during polymer-analogous reactions should be avoided or at least reduced. Comparable block copolymers have attracted interest for metal complexation [89].

The conversion of a primary amino group of a PEG derivative into a tertiary alcohol in order to obtain an acid-labile protecting group on this hydrophilic carrier was described by Anzinger and Mutter [76]:

$$\text{CH}_3\text{O–POE–OH} \xrightarrow[\text{2) } Br-CH_2-COOC_2H_5]{\text{1) Naphthalene–Na}}$$

$$\text{CH}_3\text{O–POE–O–CH}_2\text{–COOC}_2\text{H}_5 \xrightarrow{CH_3MgI}$$

$$\text{CH}_3\text{O–POE–O–CH}_2\text{–}\underset{\underset{CH_3}{|}}{\overset{\overset{CH_3}{|}}{C}}\text{–OX}$$

X = H

For both reaction steps using PEG with $\bar{M} = 2000$ g mol^{-1} quantitative yields, as proved by NMR, IR, and thin layer chromatography, were reported. A series of similar derivatives were proposed in the same paper as soluble supports of which the chain ends contain the following groups:

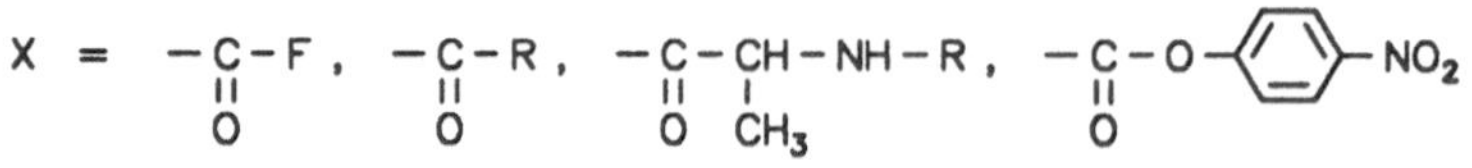

$$X = \quad -\underset{\underset{O}{\|}}{C}-F\,, \quad -\underset{\underset{O}{\|}}{C}-R\,, \quad -\underset{\underset{O}{\|}}{C}-\underset{\underset{CH_3}{|}}{CH}-NH-R\,, \quad -\underset{\underset{O}{\|}}{C}-O-\!\!\left\langle\!\!\bigcirc\!\!\right\rangle\!\!-NO_2$$

3.2.4 Special Anchor Derivatives

An interesting approach for the crucial step of the cleavage of the peptide from the polymer support was published by Tjoeng et al. [79]. They used a special polymeric protecting group which can be cleaved by photolysis and therefore under different conditions than the protecting groups of the amino acid residues.

This method was illustrated by using 3-nitro-4-bromo-methylbenzyl poly(oxyethylene) as support for liquid-phase synthesis [79]:

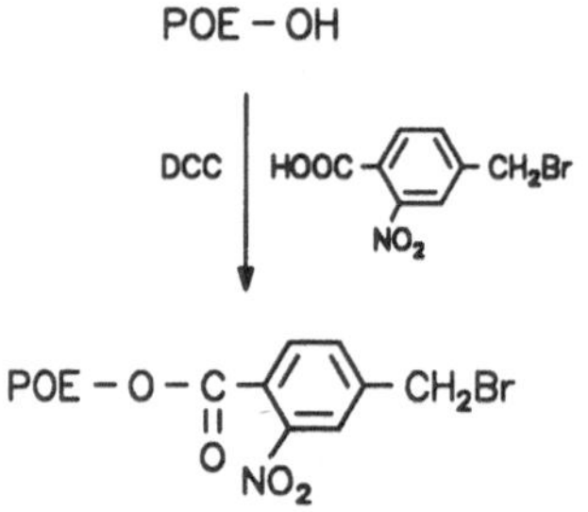

The amino acids were coupled to this support ($\bar{M} = 6000 \, \text{g mol}^{-1}$; $0.2 \, \text{mmol Br g}^{-1}$) in a repetitive manner and then the peptide was cleaved by irradiation at a wavelength above 350 nm. They found 98% yield for photolytic cleavage from the soluble polymer and, for comparison, 69% from crosslinked polystyrene.

In another study Tjoeng et al. [80] showed successfully that low yields of cleaved peptide obtained by the usual methods can be significantly improved in liquid-phase synthesis by employing this type of photocleavable polymer support.

Similarly structured derivatives of poly(oxyethylene) for preparation of easily removable soluble supports were investigated by Pillai et al. [78, 90]. They prepared acid-cleavable and photocleavable derivatives by reaction of PEG with aliphatic and aromatic isocyanates and showed their applicability by synthesizing several model peptides. The most promising examples reported are the following which have been used successfully for the synthesis of a tetrapeptide: Cleavage of the peptides from the polymer was studied in detail to establish optimum conditions for these types of supports with $\bar{M} = 6000 \, \text{g mol}^{-1}$ and average loading capacities of $0.050–0.054 \, \text{mmol g}^{-1}$.

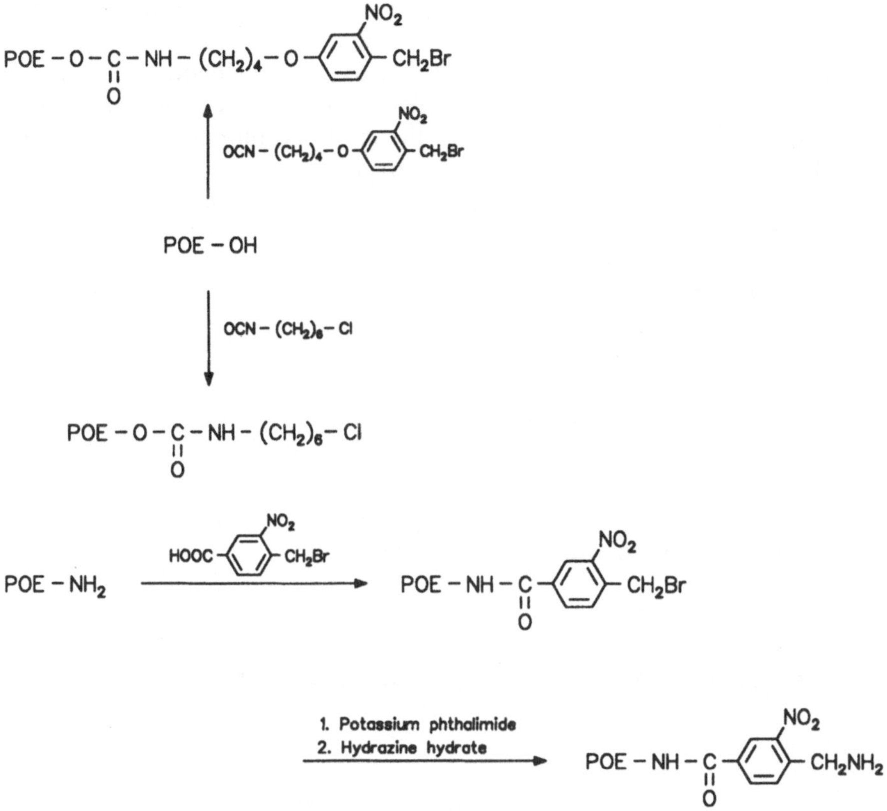

POE $-$ O $-$ C $-$ NH $-$ (CH$_2$)$_4$ $-$ O $-$ ⟨NO$_2$⟩ $-$ CH$_2$Br
‖
O

↑ OCN $-$ (CH$_2$)$_4$ $-$ O $-$ ⟨NO$_2$⟩ $-$ CH$_2$Br

POE $-$ OH

↓ OCN $-$ (CH$_2$)$_6$ $-$ Cl

POE $-$ O $-$ C $-$ NH $-$ (CH$_2$)$_6$ $-$ Cl
‖
O

POE $-$ NH$_2$ →[HOOC $-$ ⟨NO$_2$⟩ $-$ CH$_2$Br] POE $-$ NH $-$ C $-$ ⟨NO$_2$⟩ $-$ CH$_2$Br
‖
O

→ [1. Potassium phthalimide / 2. Hydrazine hydrate] POE $-$ NH $-$ C $-$ ⟨NO$_2$⟩ $-$ CH$_2$NH$_2$
‖
O

Specially functionalized soluble polymer supports were described by Colombo [82, 91]. He used monofunctional PEG ($\bar{M}$ = 5000 g mol^{-1}) for the introduction of a group which allows an easy cleavage from the polymer after synthesis of the peptide:

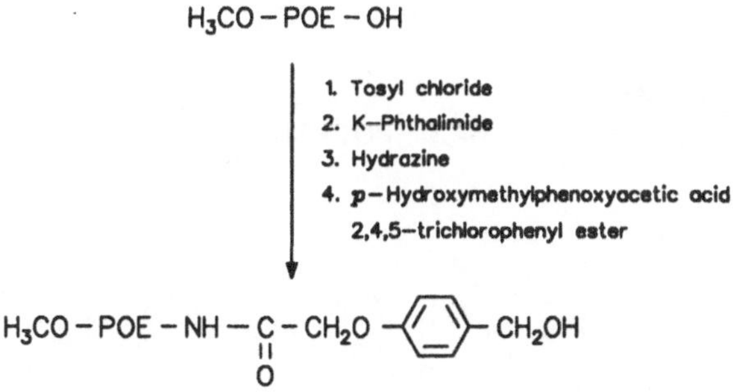

H$_3$CO $-$ POE $-$ OH

↓ 1. Tosyl chloride
2. K$-$Phthalimide
3. Hydrazine
4. $p-$Hydroxymethylphenoxyacetic acid
   2,4,5$-$trichlorophenyl ester

H$_3$CO $-$ POE $-$ NH $-$ C $-$ CH$_2$O $-$ ⟨ ⟩ $-$ CH$_2$OH
‖
O

The attached peptides can be cleaved from this soluble carrier by anhydrous trifluoroacetic acid whereas a transformation of the hydroxybenzyl to the aminobenzyl group and the additional introduction of a nitro group in the 3-position permits photolytic cleavage [91].

Colombo introduced additional interesting derivatives of poly(oxyethylene) as polymer supports in liquid-phase peptide synthesis [77]. Specially designed for fragment condensation, he investigated the oxycarbonylhydrazide group for the synthesis of peptide hydrazides. Starting from the α-amino-ω-methoxy poly(oxyethylene), which he obtained *via* the tosyl and phthalimido derivative, he used the 4-nitrophenyl ester of 4-acetyloxymethylbenzoic acid and phenyl chloroformate as reactants for the following reaction sequence:

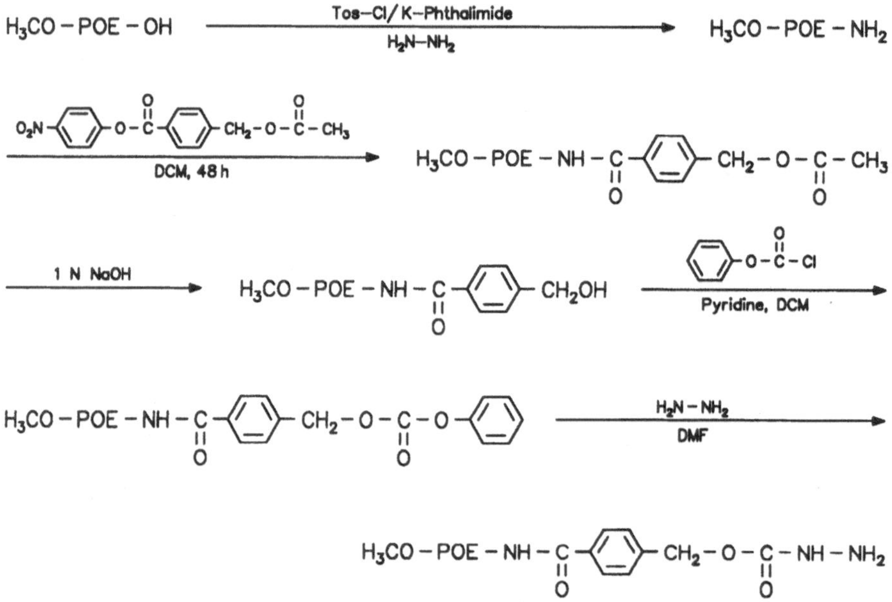

Another route of synthesis was described for the preparation of a 4-alkoxybenzyloxycarbonyl hydrazide polymer [77]:

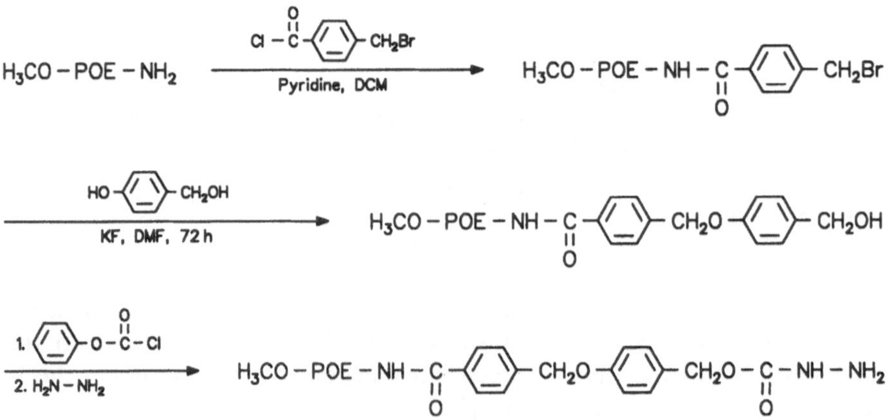

The polymer supports were purified by precipitation–crystallization from dichloromethane and dimethylformamide with diethyl ether. The overall yields of the peptide hydrazides synthesized (up to pentapeptides) with this hydrophilic support were in the range between 30 and 53%. Rates of cleavage from the support either by catalytic hydrogenolysis or by trifluoroacetic acid in DCM were reported to be very high (90–96%).

Recently, a new PEG derivative was reported by Tjoeng and Heavner [81]. They synthesized 4-(2-chloropropionyl)phenylacetic acid in three steps for terminal functionalization of PEG ($\bar{M}$ = 6000 g mol^{-1}; 0.33 mmol g^{-1}) in order to introduce a photolabile group:

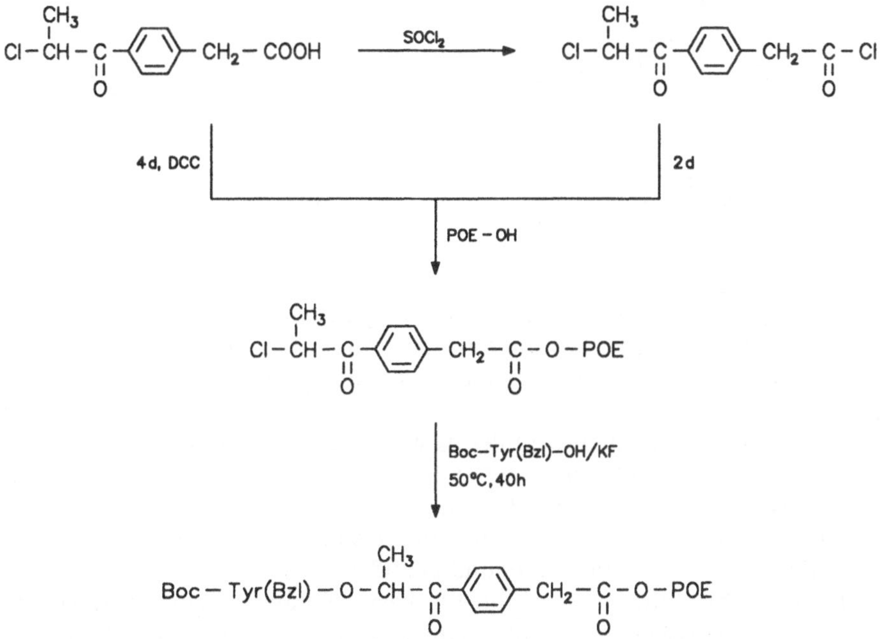

The chloroderivative of POE was obtained in 64% yield (0.21 mmol Cl g^{-1}), whereas the first amino acid (Tyr) was attached with 86% yield referred to Cl sites. The final functional capacity was 0.18 mmol g^{-1} for Tyr corresponding to a total loading capacity of about 50%. The practical use of this carrier was demonstrated by the synthesis of the protected pentapeptide Z–Arg(Z,Z) –Lys(Z)–Asp(OBzl)–Val–Tyr(Bzl)–OH, the active segment of the bovine thymus peptide Thymopoietin. The photolytic cleavage of the protected penta-peptide was carried out successfully (350 nm; 18 h) and gave 330 mg product (92% yield). Subsequent acid hydrolysis of the support for control revealed that no detectable peptide remained on the carrier, thus providing quantitative photocleavage.

3.2.5 Functionalization for Cysteine Entrapping

A polymer support specially designed for cysteine-containing peptides was suggested by Glass et al [92]. They prepared a support from PEG and 4-phenoxy-3,5-dinitro-benzoyl chloride with a capacity of 0.1–0.2 mmol g^{-1}. This soluble support reacts rapidly and selectively with the sidechain sulfhydryl groups of cysteine in aqueous buffers at pH 7. Thus, polymer-bound peptides are obtained with a thiolsensitive dinitrophenylene link which permits an easy release of peptides from the polymer:

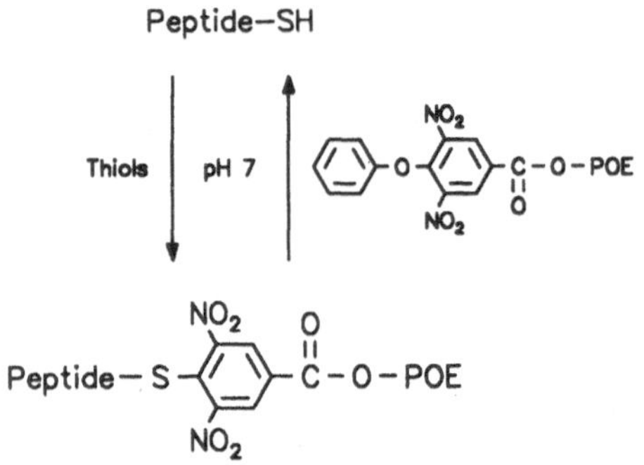

Peptide—SH

For example, cleavage was accomplished by addition of 2-mercaptoethanol in aqueous solution at pH 7 within one hour. Another interesting aspect of this approach is that prior masking of other functional groups of the peptide is not required.

Polymer-supported small peptides as well as the B-chain of bovine insulin were reported to be soluble in water and in some organic solvents when using PEG derivatives with a molecular mass of 6000 g mol^{-1} to this purpose. Enzymatic investigations showed that this linear soluble support does not restrict the access of enzymes to bound peptides.

Polymer-bound z-nitrostyrenes on the basis of polystyrene/divinylbenzene and of poly(oxyethylene) represent polymer products useful for addition reactions with the thiol groups of cysteine peptides or any other thiols in aqueous media. Thus, cysteine peptides can be removed selectively by reversible addition from peptides mixtures.

Soluble supports ($\bar{M} = 6000$ g mol^{-1}) of the following structure have been used with 4-(2-nitrovinyl)benzoic acid as terminal group (loading capacity: 86%) [93]. The loading capacity of the latter POE-derivative was 75% (0.25 mmol g^{-1}). Such solubilizing polymer-based protecting groups are of considerable interest for the synthesis of cysteine peptides.

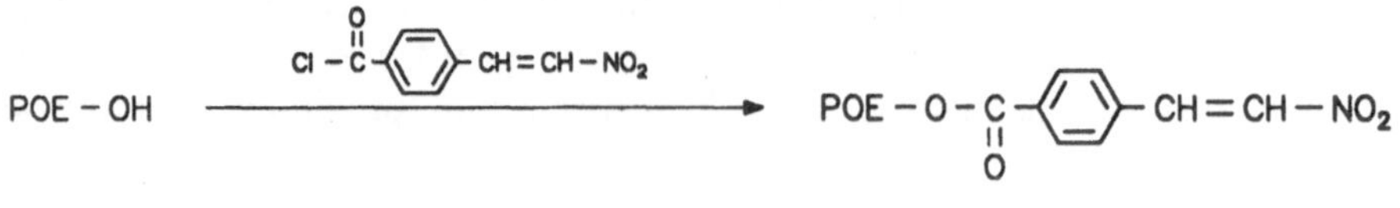

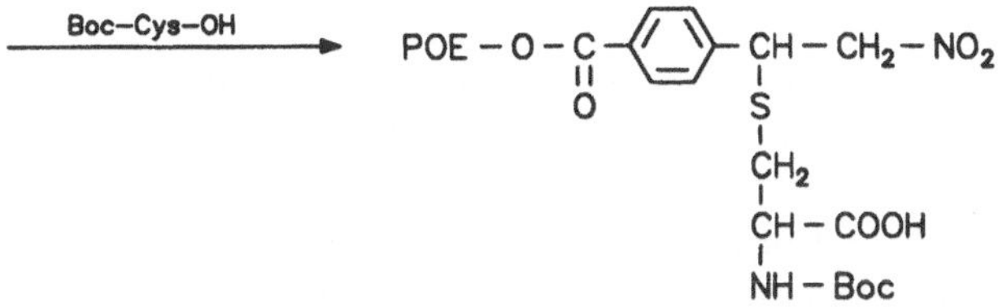

Boc–Cys–OH $\longrightarrow$ POE $-$ O $-$ C $\langle$⬡$\rangle$ $-$ CH $-$ CH$_2$ $-$ NO$_2$

with the C attached to O (C=O), and below CH: S, CH$_2$, CH $-$ COOH, NH $-$ Boc

3.3 Soluble Polystyrene as Carrier

The Merrifield idea of simplifying peptide synthesis by attaching one end to a support was first transferred to soluble polymers by Shemyakin et al. [94] and Ovchinnikov et al. [11]. They tried to overcome the problems in solid-phase synthesis by using the same support but in a soluble, non-crosslinked form. The polystyrene ($\bar{M}$ = 200 000 g mol^{-1}) prepared by emulsion polymerization had 25% chloromethylated benzene rings resulting in a loading capacity of 0.91 mmol g^{-1} (glycine). After the initial attachment of the amino acid and also after each synthetic step they removed excess reagents by precipitation into water. Although this simple procedure was satisfactory for the preparation of the tetrapeptide Gly–Gly–Leu–Gly, syntheses of other peptides containing heterocyclic amino acids, for example, revealed the critical points of this support and procedure. Precipitation is involved with some essential problems, especially when a hydrophobic carrier such as polystyrene is precipitated with water. Also, the high requirements for this support in terms of structural homogeneity and solubilizing power are not fulfilled.

Some years later, Green and Garson reported another attempt to improve the Merrifield method by the liquid-phase technique [57]. They successfully synthesized a pentapeptide using chloromethylated polystyrene ($\bar{M}$ = 230 000 –250 000 g mol^{-1}) which was desulfurized prior to the functionalization reaction in order to remove sulfur-containing free-radical terminating agents. Beside investigations on the optimization of polymer-analogous chloromethylation of polystyrene, they also studied the attachment of N-protected alanine to the support in detail. By infrared monitoring of this fixation reaction it was found that the maximum yield was obtained only after about 80 h at room temperature:

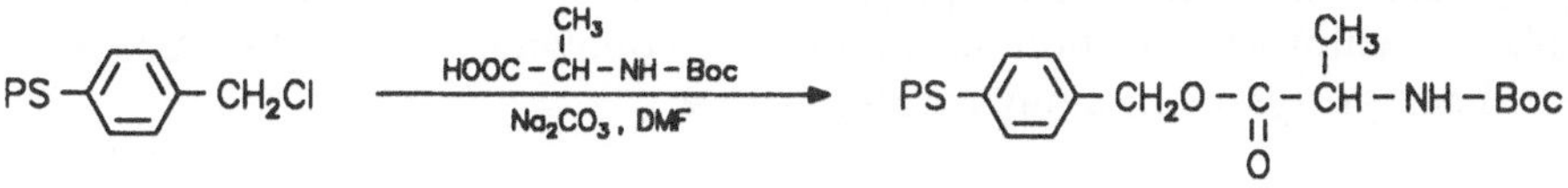

PS $\langle$⬡$\rangle$ $-$ CH$_2$Cl $\xrightarrow[\text{Na}_2\text{CO}_3,\ \text{DMF}]{\text{HOOC} - \overset{\text{CH}_3}{\underset{}{\text{CH}}} - \text{NH} - \text{Boc}}$ PS $\langle$⬡$\rangle$ $-$ CH$_2$O $-$ C $-$ $\overset{\text{CH}_3}{\underset{}{\text{CH}}}$ $-$ NH $-$ Boc (with C=O)

The optimum loading capacity was suggested as being about 0.5 mmol g^{-1} because a higher degree of substitution would not be favorable by reason of a reduced solubility of the polymer-supported peptide. Noteworthy are the refinements on the purification procedure of the polymer support. According to their findings, precipitation is best accomplished by pouring a 10% solution of the polymer in dimethylformamide into an aqueous sodium chloride solution.

For cleavage of the peptide from the polymer three methods were suggested: (1) Hydrogen bromide in trifluoroacetic acid with the suspended polymer support, (2) hydrogen bromide in a benzene solution of the carrier, and (3) anhydrous hydrazine in dimethylformamide with heating. The method was illustrated by the synthesis of Ala-Ile-Arg-Ser-Ala and the authors emphasized the much greater flexibility of the liquid-phase method in terms of the coupling techniques which can be used.

Early attempts to improve solid-phase synthesis were also made by Maher et al. [95, 96]. In order to circumvent some disadvantages of crosslinked polymer supports they prepared chloromethylated polystyrene without crosslinking but with a relatively high degree of substitution of 62%. The unfunctionalized polymer was soluble in benzene but the Boc-Gly-support was reported to be insoluble in DMF and chloroform.

The same support, but a completely different purification technique, was suggested by Andreatta and Rink a few years later [24]. They used relatively low-molecular weight polystyrene ($\bar{M} = 20\,400 \, \text{g mol}^{-1}$) and gelfiltration for removal of the excess reagents. In addition, the chloromethylation of the basis support and the attachment step were studied intensively. A content of 10–30% of chloromethyl groups remained since no quantitative conversion during the fixation reaction of the first amino acid could be obtained. Cleavage from the support was reported to be 90% if hydrogen fluoride at 0 °C or hydrogen bromide in trifluoroacetic acid were used. Interestingly, they also studied the limits of the loading capacity for the solvent dichloromethane which served as solvent exclusively. Pointing to the dependence of solubility on the type of amino acid, sequence of peptide, and type of protecting groups in general, they found a capacity of 0.5 mmol g^{-1} to be without any problems during synthesis in the case of PS–Phe–Pro–His(Bzl)–Val–Tyr(Bzl)–Val–Boc. Strong restrictions are met by using gelfiltration as the purification method; precipitation was not investigated. Remarkable at that time were the kinetic studies of polymer-analogous reactions for both linear and crosslinked polystyrene. As a consequence of these results, the introduction of a monovalent, soluble polymer support was proposed [24].

One of the few studies on the application of peptide fragment condensation on soluble polymer supports was published by Narita et al. [39, 97–100]. First, they prepared the chloromethyl polystyrene support, not by the usual polymer-analogous chloromethylation reaction but by copolymerization of styrene with chloromethyl styrene in a molar ratio of 1 : 9:

The chloro-functional comonomer was a mixture of *m*- and *p*-isomers (7 : 3) but the position of substitution seemed not to influence polymer-analogous reactions essentially when used in the form of a soluble polymer support. The chlorine content was 0.91 mmol g^{-1}, as expected.

As basis support, PS–CH$_2$O–Leu–H was used with a loading capacity of 0.62 mmol g^{-1} to which the tetrapeptide Boc–Tyr(Bzl)–Gly–Gly–Phe–OH was coupled in DMF as solvent within 5 h. Excess reagents were removed by precipitation into methanol and the recovery of the polymer-supported peptide was higher than 93% on the basis of yields. Interestingly, they found no influence of the peptide chain length on the reactivity of the C-terminal amino acid in the coupling reaction using DCC and other reagents. They stated that shortcomings of the solid-phase and the classical peptide synthesis are eliminated by the use of soluble polymer supports.

Recently, an alternative to the crosslinked polystyrene support was suggested by Sheppard but still using the solid-phase technique [101]. Very recently, Isokawa et al. [102] presented an interesting study on fragment condensation of peptides using soluble polymer supports. For the basis polymer polystyrene, they compared the advantages and disadvantages of both liquid-phase and solid-phase peptide synthesis. For example, the coupling rates of Boc-oligopeptides (*n* = 3–10) relative to those of Boc–Val–OH were determined by competitive coupling reactions with amino acids attached to soluble and crosslinked polystyrene supports using four different coupling systems. Also the influence of the peptide chain lengths of Boc-oligo-peptides and the degree of crosslinking of polymer supports on the coupling rates of the C-terminal amino acids for the coupling reactions with free terminal amino groups of amino acids attached to polymer supports were studied. Soluble polystyrene used in this study contained 1 mol % of amino acid per styrene monomer unit and crosslinked polystyrene 1 or 10 mol %.

Having initiated the study because of the serious problems in fragment condensation on crosslinked polymer supports, they actually found that amino groups were present in the support poly(styrene–*co*–divinylbenzene) which were sterically inaccessible for peptide fragments such as Boc–(Ala)$_2$–(Leu)$_3$–OH and Boc–(Ala)$_2$–(Leu)$_3$–(Pro)$_2$–(Leu)$_3$–OH. Finally, they emphasized the versatility of soluble supports and the limitations of crosslinked polymer supports for peptide fragment condensation.

3.4 Miscellaneous Soluble Polymer Supports

A quite different approach with respect to the support backbone was taken by Blecher and Pfaender using poly(ethyleneimine) as water-soluble support [37, 38, 103]. They synthesized the model peptide Ala–Trp–Ile–Arg in aqueous phase by application of the *N*–carboxyanhydride method. The peptide was attached through an arginine as anchoring group for the peptide to the polymer chain:

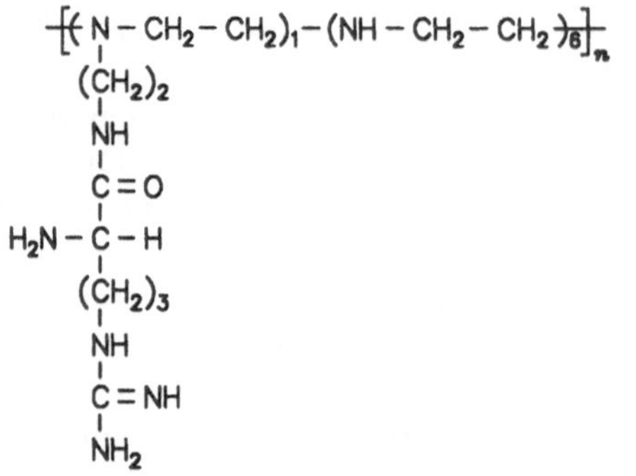

$$\begin{array}{l}
\text{---}\!\!\big(\!\!\big(\,N\!-\!CH_2\!-\!CH_2)_1\!-\!(NH\!-\!CH_2\!-\!CH_2\,\big)_8\big)_n\\[2pt]
\qquad\ \ (CH_2)_2\\
\qquad\ \ \ \ NH\\
\qquad\ \ \ \ C\!=\!O\\
H_2N\!-\!C\!-\!H\\
\qquad\ \ (CH_2)_3\\
\qquad\ \ \ \ NH\\
\qquad\ \ \ \ C\!=\!NH\\
\qquad\ \ \ \ NH_2
\end{array}$$

Noteworthy is the high loading capacity of this carrier: the poly(ethyleneimine) support $(30\,000 < \bar{M}/(\text{g mol}^{-1}) < 40\,000)$ contained 18 mol% arginine which corresponds to a capacity of 2.25 mmol Arg g^{-1} support. It could be even raised up to a limit of a molar ratio 1:3 (Arg:monomer unit) without losing solubility in water. Also remarkable is the cleavage of the peptide from the polymeric support which was carried out quantitatively in 2 h by enzymatic cleavage with trypsin.

A support prepared by copolymerization of 1-vinyl-2-pyrrolidinone and vinyl acetate was introduced by Bayer and Geckeler [15]. By subsequent polymer-analogous saponification they obtained a support with hydroxyl groups which is readily soluble in water and various organic solvents:

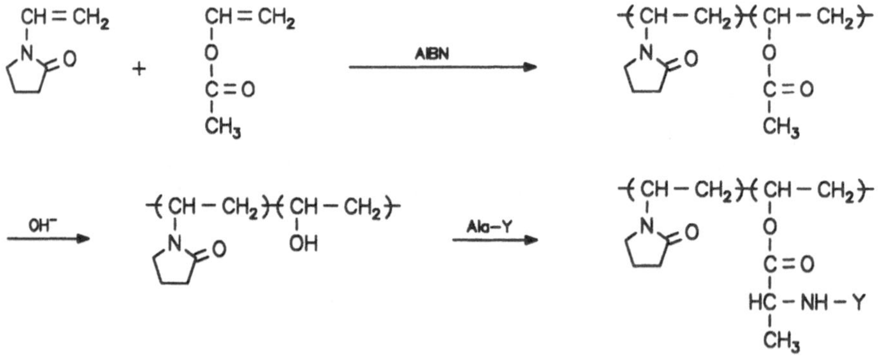

The average molecular mass was 33 000 g mol^{-1} and the polymer was purified by precipitation into cyclohexane and subsequent ultrafiltration. Symmetric anhydrides were used as coupling reagents for the synthesis of a tetrapeptide during which the polymer-supported peptide chain remained completely soluble. Therefore, membrane filtration was found to be effective for purification of the polymer support. This carrier seems mainly applicable for smaller oligopeptides because of the limited solubilizing power of the heterocyclic comonomer units.

Another new route of synthesis for soluble polymer supports based not on the usual polymer-analogous derivatization but on the preparation of the support by copolymerization of functional monomers was outlined and demonstrated by Geckeler and Bayer [73, 104]. They used radical copolymerization of amino acid alkenyl esters, which are easily accessible by a vinyl exchange reaction with vinyl acetate, with 1-vinyl-2-pyrrolidinone as solubilizing co-monomer for direct preparation of a functional soluble support [104]. Thus, a carrier soluble in water as well as in many organic solvents was obtained which already contained the first amino acid residue of the peptide:

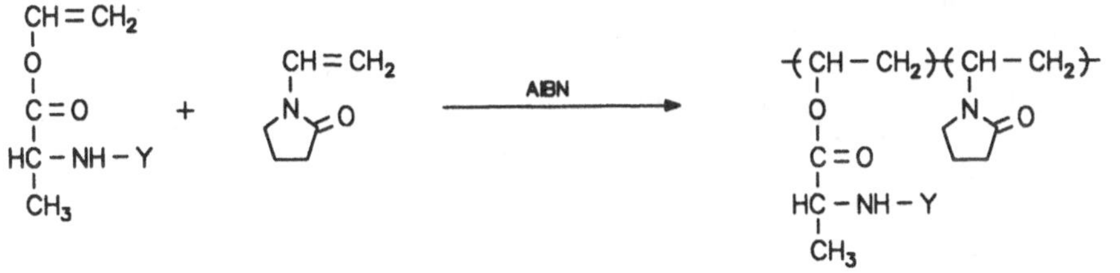

After cleavage of the amino protecting group, the next amino acid of the sequence was coupled directly to the support. For coupling DCC was used and the ultrafiltration technique for removal of excess reagents from the polymer support.

3.5 Combined Use of Solid and Soluble Supports

3.5.1 Alternating Application

Frank and Hagenmaier suggested an alternative to both solid- and liquid-phase peptide synthesis by creating the "solid-liquid-phase" method in order to overcome the problem of failure and truncated sequences [105, 106]. To this end, poly(ethylene glycol) monoalkyl ethers were used as carboxy protecting groups. The principle of this method is depicted in Scheme 2.

It is characterized by temporary immobilization of the peptide chain during step-wise coupling reactions which provides the experimental ease of the solid-phase technique. In addition, fixation to the polymer allows the use of a high excess of reagents that helps to avoid truncated sequences and to obtain very pure products. It is apparent, that according to this method, the peptide grows on the soluble polymer but coupling is carried out with matrix-bound carboxyl components. The advantages of this approach could be reduced, however, by the problems which should be expected by heterogeneous polymer–polymer reactions.

This method, which facilitates the separation of reacted and unreacted amino components at the end of each coupling step, was exemplified by the synthesis of the pentapeptide Gly–Val–Gly–Ala–Pro. The same peptide was

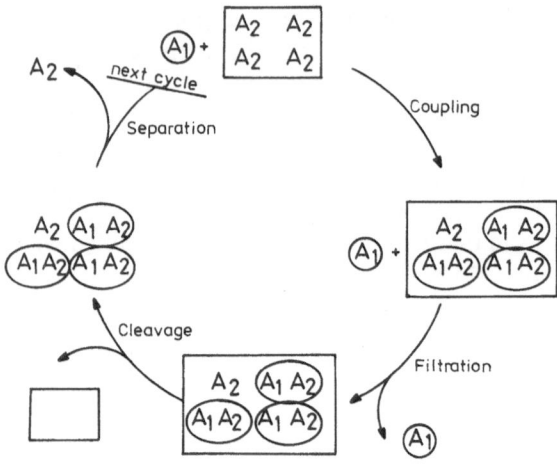

A$_1$, A$_2$ = Amino acid(residue), peptide chain

[ ] = Solid support, amino protecting group

○ = Carboxyl protecting group

Scheme 2. Scheme of the alternating solid-liquid-phase method for peptide synthesis [105]

synthesized with the monostearylether of PEG as carboxyl protecting group [107]. Using this method the free amino acid was separated by precipitation with ethanol or by membrane filtration.

Another interesting approach combining the advantages of both solid- and liquid-phase peptide synthesis was proposed by Jung et al. [108–110] (Fig. 5). This method is based on the reaction of an excess of active esters attached to an insoluble, crosslinked polystyrene matrix with soluble PEG-supported C-terminal amino acids. Thereby, repetitive sequential coupling steps are carried out on the soluble support (M = 6000 and 10000 g mol^{-1}, loading capacity of 78 and 56%) whereas the matrix-bound carboxyl component serves for coupling. This allows one to remove excess reagents by rapid and convenient filtration after each coupling step as done in solid-phase synthesis [108]. Surprisingly, the polymer-polymer reactions proceed satisfactorily without being hampered essentially by steric factors. Although several tripeptides have already been successfully synthesized, a widespread application of this procedure will need the proof of synthetic reliability for preparation of other sequences and longer oligopeptides.

3.5.2 Covalent Combination by Grafting

Recently, attempts were made to combine the advantages of both the solid-phase and the liquid-phase method of repetitive-type synthesis by a "mixed"

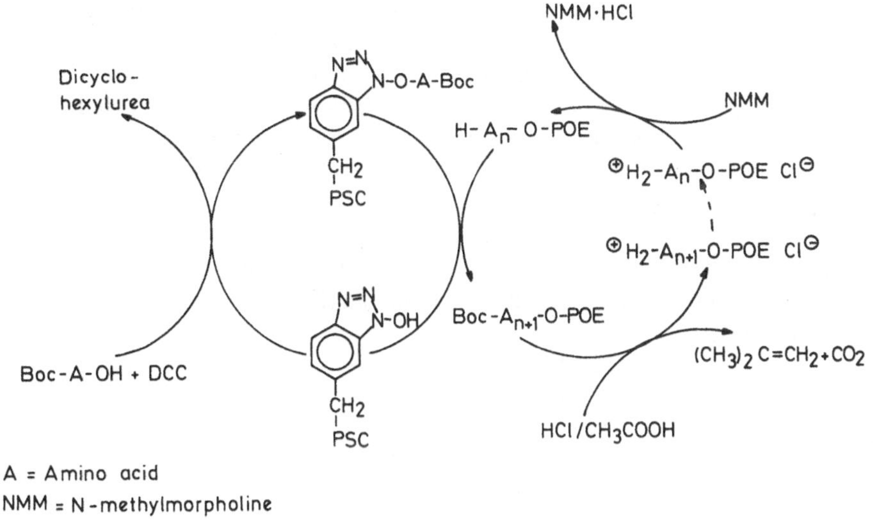

A = Amino acid
NMM = N-methylmorpholine

Scheme 3. Peptide synthesis using a combination of soluble peptide polymers and polymer reagents [109, 110]

support. To this end, poly(ethylene glycol) chains were grafted onto function-alized crosslinked polystyrene resins in order to obtain immobilized hydrophilic supports. This type of carrier facilitates both the homogeneous solvation of the reaction site and the insolubilization of the substrate for easy work-up. The favorable swelling characteristics in many solvents lead to kinetics of the coupling reactions which proved to be linear thus showing the equivalence of all functional groups.

Because such polyether spacer polymers are also of interest in other fields of polymer-supported immobilization, the first reports published on the synthesis of polystyrene graft copolymers date back to almost one decade ago [111, 112]. In fact, these graft supports have been applied successfully in fields where insoluble polymers with homogeneous phase reactions are advantageous, e.g., for phase-transfer reactions [113–116].

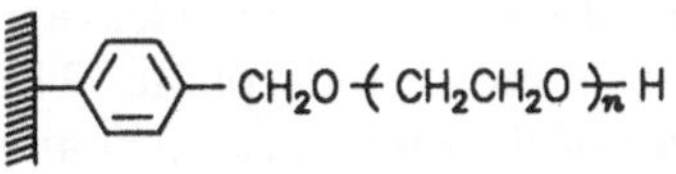

This type of polymer support seems to be an interesting alternative to conventional polystyrene because the simple procedures for the experimental work-up of the Merrifield synthesis can be applied and simultaneously quasi-homogeneous reaction conditions are attained by reason of the hydrophilic poly(oxyethylene) spacer group.

For example, by introduction of selectively cleavable anchoring groups a temporary immobilization of the polymer-bound peptides could be attained

[117, 118]. The functional capacities were up to 1 mmol g^{-1} and the molecular masses of the hydrophilic spacer between 400 and 2000 g mol^{-1}. As a typical example, the tetrapeptide Ile–Ala–Val–Gly was synthesized with this type of polymeric support. Kinetic investigations on the aminolysis of activated amino acids with glycine esters for these graft copolymers showed that the reactivity of the terminal amino group is of the same order as that in homogeneous reactions.

$$\text{PS–CH}_2\text{–X–POE–OH}$$

PS = insoluble polymeric matrix: crosslinked polystyrene
POE = hydrophilic spacer: poly(oxyethylene) chain
X = anchoring group: –O–
                     –CO–O–
                     –CO–CH$_2$–NHCO–O–

4 Soluble Supports for the Synthesis of Oligonucleotides

4.1 General Aspects

4.1.1 Introduction to Liquid-Phase Nucleotide Synthesis

The second domain, in which the liquid-phase method has been applied after the introduction into peptide synthesis, is the oligonucleotide synthesis [119–123]. Only a few years after the first reports on successful liquid-phase synthesis of peptides, the first attempts to apply this method to the complicated synthesis of oligonucleotides were made [122–130]. The period of this time interval is similar to that in solid-phase synthesis. Liquid-phase oligonucleotide synthesis is, comparable to peptide synthesis, advantageous for several reasons: First of all it should be mentioned that all reactions, coupling of nucleotides and cleavage of protecting groups, can be carried out in the homogeneous phase. The polymeric protecting group increases the solubility of the oligonucleotide and eases separation and purification from by-products. In addition, this method permits easy analytical control during synthesis and also intermediate purification if necessary.

The method appears especially attractive for the synthesis of lower oligonucleotides, although hexanucleotides have already been synthesized. An important domain is fragment condensation or its combination with other methods.

The rapid decrease of yields of the individual reaction steps is not as important as in solid-phase synthesis because products can be purified in between or coupling reactions can be repeated, if necessary. Thus, it is possible to reduce the formation of failure and truncated sequences by eliminating the diffusion problems of the reactants in the polymer matrix based on steric hindrance and reduced accessibility of the reagents.

The basic requirements for a suitable soluble polymer support in oligo-nucleotide synthesis are mechanical stability, chemical inertness, and a high solubilizing power. The solubility properties of the polymer carrier are especially important in this field because of the relatively large size and high polarity of oligonucleotides.

The application of soluble polymers as supports in this field has essential advantages in comparison to the solid-phase technique or conventional con-densation in solution [28, 29, 31, 36, 40]:

- The oligonucleotide chain is attached to a carrier during synthesis; thus, problems based on solubility differences of the monomers are of minor importance.
- Separation and purification in the synthesis cycle are reduced to precipitation or membrane filtration which offers an easy and rapid procedure.
- The principle of the method permits the use of a high excess and repeated coupling of reagents so that optimum yields of the coupling steps are easier to obtain.

However, because of the complex nature of nucleotide coupling reactions and non-optimum yields of the desired products, the synthesis of higher oligonucleotides or polynucleotides seems to be problematic so far [131–133].

4.1.2 Soluble Supports in the Synthesis Cycle

Comparable to peptide synthesis, the liquid-phase synthesis of oligonucleotides is based on the sequential repetition of reaction steps, thus adding one mono-meric nucleotide to the other (Scheme 4).

The first nucleotide is attached to the functional soluble polymer *via* the phosphate group. After cleavage of the protecting group, reagents and other low-molecular compounds are separated by membrane filtration. Then, the product can be controlled analytically and the next nucleotide units are coupled repeatedly in the same cycle until the desired length of the oligonucleotide is attained. Finally, the product is cleaved chemically or enzymatically from the polymer support.

The crucial steps in terms of the soluble polymer support are the following:
- Availability of an adequate basis support or an appropriate functionalization reaction,
- attachment of the first nucleotide of the projected sequence to the carrier,
- blocking of unreacted functional groups of the polymer,
- cleavage of the oligonucleotide from the support, and
- separation of the product from the polymeric carrier.

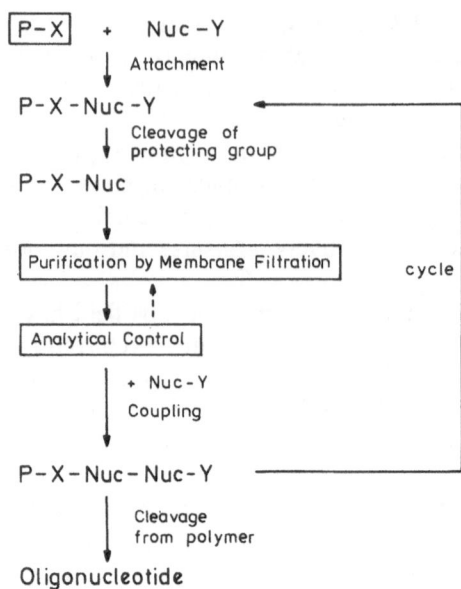

P = polymer
X = functional group
Nuc = nucleotide
Y = protecting group

Scheme 4. Scheme of liquid-phase oligonucleotide synthesis

4.1.3 Attachment and Cleavage Reactions

Generally, two different strategies are used in conventional synthesis of oligo-nucleotides: the "diester" and the "triester" method [131, 132]. In syntheses with soluble polymer supports and in recent works in general the "diester" method is preferred. Thereby, the 5'-O position is protected by the soluble polymer and the elongation of the oligonucleotide chain is carried out *via* the 3'-O position. Most researchers used an ether, ester, or a carbonate bond for the 5'-nucleoside linkage.

Apart from basis polymers, such as poly(vinyl alcohol), which contain already the functional group for attaching the first nucleoside, the polymer has to be derivatized in many cases in order to obtain the suitably functionalized support [31, 126, 134, 135]. For the selection of the appropriate functional group, one has to consider that the bond between the support and the oligonucleotide must be resistant to cleavage during the repetitive synthesis. In addition, the same bond should be cleavable at the end of synthesis without affecting the internucleotide bonds. Obviously, the cleavage of the oligonucleotide product after finished synthesis is mainly dependent on the type of the polymer-nucleotide bond. Therefore, a variety of cleavage conditions were used for the different types of linking. Typical examples are given in Table 12.

The acid-labile trityl ether, for example, is very easy to cleave off without interference of other protecting groups during synthesis. An interesting aspect of protecting and cleaving techniques is the introduction of a thioether as safety-catch moiety which is eliminated by oxidation to the sulfone [134].

As usual in repetitive-type, polymer-supported synthesis, the product is a mixture of the desired oligonucleotide and corresponding shorter homologues. Separation and purification of the products by chromatographic methods is therefore a must [60].

4.1.4 Selection and Survey of Soluble Supports

The selection of the polymer support is the most important step in liquid-phase synthesis of oligonucleotides. The number of soluble polymers, which could serve here as carriers, is very limited due to many restrictions. In most cases synthetic macromolecules are preferred because they fulfill the requirements better than many biopolymers.

A series of different polymers have been described for this purpose, however, it seems that some basis polymers are strongly preferred. Among these are mainly polymers containing backbones of polystyrene or poly(oxyethylene). In general, many of these supports provide condensation yields similar to those in conventional, non-supported synthesis in the homogeneous phase.

In the case of oligonucleotide synthesis two solvents are mainly used: pyridine and aqueous media. Therefore, hydrophobic carriers were used only in

Table 12. Examples for the type of bond between polymer support and nucleotide and typical reagents for their cleavage

Type of bond of polymer-nucleotide	Reagents for cleavage from the support	Ref.
Trityl ether	Trifluoroacetic acid in chloroform (1%)	59
Trityl ether	Trifluoroacetic acid in dioxane (50%)	40
Trityl ether	–	31
Phosphodiester	N-Chlorosuccinimide/0.1 M phosphate buffer; 1 N sodium hydroxide solution	134
Phosphodiester	–	31
Phosphoramidate	Isoamyl nitrite in acetic acid/pyridine	29
	Acetic acid (80%)	40
Carbonate	Concentrated ammonium hydroxide solution	136

the beginning and the search for polymer supports was directed later towards hydrophilic macromolecules [29, 31, 36, 40]. Data with respect to polymer supports, which have been presented in leading journals, are summarized in Table 13.

The majority of the studies in this field stress the synthesis of oligonucleotides and not the polymer support. An optimization of the factors in the application of soluble carriers needs more intensive investigation because of the complex interaction of many parameters during polymer-supported reactions.

4.2 Soluble Polystyrene Supports

In addition to the insoluble matrix polymers, which are based mainly on polystyrene crosslinked by divinylbenzene, linear-chain polystyrenes have attracted the interest of nucleotide chemists. Nevertheless, one of the main disadvantages of this carrier is the strong hydrophobicity which is unfavorable in terms of compatibility with the polar phosphodiester bonds of the nucleotide chain. These polystyrenes without crosslinks are soluble in a series of organic solvents, however, they are insoluble in water and ethanol. The average molecular masses of the polymers were in the range of up to 3.10^6 g mol^{-1} [40, 41, 59].

Very soon after the introduction of the solid-phase synthesis in peptide chemistry, Cramer et al. [40] published a paper on oligonucleotide synthesis with soluble polystyrene as support. They used polystyrene ($\bar{M}$ = 170 000 g mol^{-1}) with 20% of monomethoxytrityl- functionalized phenyl groups of which 40% could be loaded with 3'-O-acetyl thymidine (Scheme 5).

Table 13. Soluble polymer supports for oligonucleotide synthesis

Item No.	Polymer backbone	$\bar{M}$ 10^3 g mol^{-1}	Functional group	Products synthesized	Yield (%)	Ref.
1	Polystyrene	270	Monomethoxytrityl chloride	dTpdT dTpdC dT(pdT)$_2$	96 91 83	41, 59
2	Polystyrene	170	Monomethoxytrityl chloride	dT(pdT)$_2$	11	40
3	Polystyrene	170, 600	Aminophenyl	–	–	137
4	Poly(vinyl acetate-co-1-vinyl pyrrolidinone-2)		Hydroxyl	dTpdT	56	136
5	Poly(vinyl alcohol)	70	Hydroxyl	(pdT)$_2$ (pdT)$_5$	51 8	34 48
6	Poly(oxyethylene)	20	Hydroxyl	–	–	31
7	Poly(oxyethylene)	10	Aminoethyl	d(pdA)$_3$ d(pdT)$_5$	14 8	29
8	Poly(oxyethylene)	20	Monomethoxytrityl chloride	–	–	31
9	Poly(oxyethylene)	6	2-Hydroxyethyl 4-aminophenylthioether	–	–	134

The yield of TpTpT was 11%. In the same paper they suggested non-crosslinked amino polystyrene as a soluble polymer support in combination with the acid-labile phosphoric acid amide bond for attachment to the carrier. The functional capacity of this support was in the range of 25%. In this case, 9% of the amino groups could be coupled with 5'-thymidylic acid using DCC as condensating reagent.

Simultaneously, Hayatsu and Khorana reported the same polymer being used as support for deoxyribonucleotide synthesis [41, 59]. They gave, however, more experimental details and noted much higher yields using polystyrene with an average molecular mass of $270\,000\,\mathrm{g\,mol^{-1}}$.

The functional group introduced into the polymer was the 4-methoxytrityl residue for several reasons. It is easily introducible and the degree of substitution can be controlled conveniently by varying the reaction conditions. Also, derivatives with this group are known to be soluble in anhydrous pyridine and, finally, it fits excellently into the protecting group strategy of oligonucleotide synthesis. The introduction of the methoxytrityl group was accomplished by Friedel-Crafts-arylation with benzoyl chloride yielding the benzophenone derivative with about 30% conversion. The trityl chloride was obtained by heterogeneous reaction with 4-methoxyphenyl magnesium bromide and subsequent treatment with acetyl chloride (see Scheme 5). The capacity of the trityl chloride group was $0.4\,\mathrm{mmol\,g^{-1}}$ corresponding to a value of about 6% of substituted styrene units. Also, this support with a low ketone content was investigated with regard to a quantitative conversion to trityl functions.

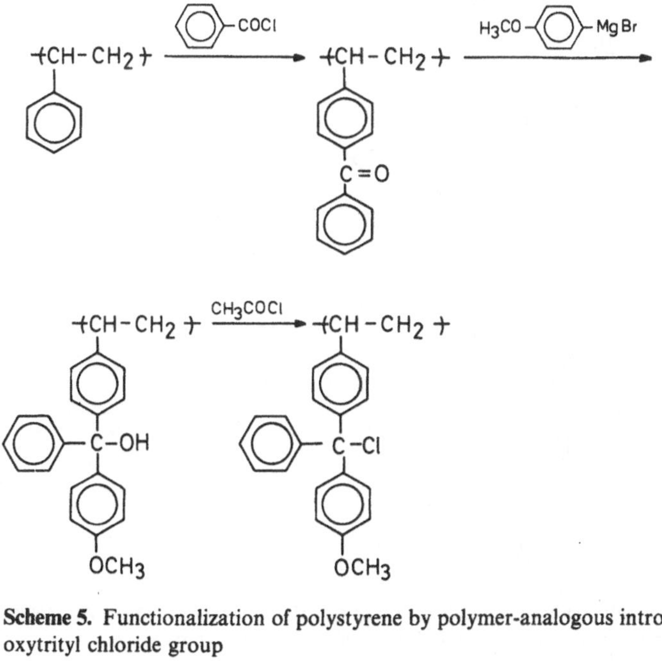

Scheme 5. Functionalization of polystyrene by polymer-analogous introduction of the monomethoxytrityl chloride group

To initiate the first step of the repetitive synthesis, the trityl chloride derivative of polystyrene was subjected to reaction with thymidine in pyridine (Scheme 6). It is note-worthy that unreacted trityl chloride groups were blocked as methyl ethers by addition of methanol [41]. The thymidine content was found to be 60–340 μmol g^{-1} after polymer-analogous cleavage from the support. This capacity corresponds to a degree of functionalization of 15–85% and points to steric problems of this attachment to the polymer.

The next condensation step was carried out by reacting the functionalized support with a 40-fold excess of 3'-O-acetyl thymidine 5'-phosphate (pT-OAc) in the presence of mesitylene-sulfonyl chloride as condensating reagent. After 2 h, the yield was already 93%. Generally, average yields obtained were 88–96% [41]. Isolation of the product was achieved by precipitation from an aqueous medium and subsequent filtration or centrifugation. Kabachnik et al. [138, 139] carried out phosphorylation reactions of nucleotides on non-crosslinked polystyrene according to Scheme 7.

Although many variations of the classical Merrifield resin have been prepared for nucleotide synthesis, most of them have not been picked up by other investigators to be studied in detail [140–149]. For example, Yip and Tsou suggested isotactic polystyrene, which is a crystalline and high-molecular weight polymer, as support for the synthesis of ribooligonucleotides [150, 151]. Modifications of which could be of great interest for liquid-phase synthesis since this support is characterized by a regular sterically defined polymer backbone.

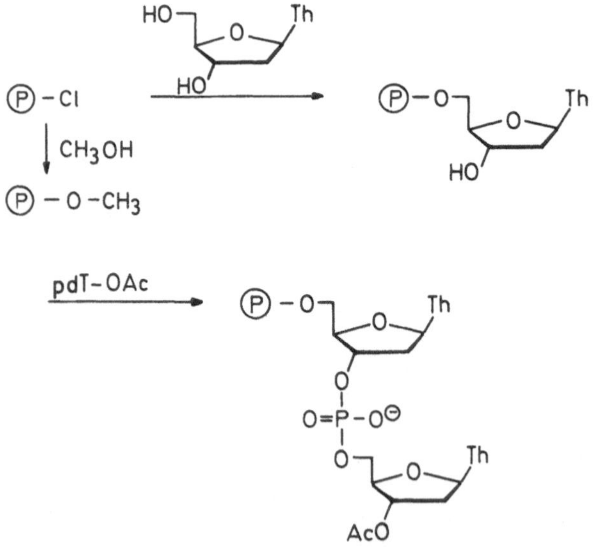

Scheme 6. Starting of the repetitive oligonucleotide synthesis by attachment of the initial nucleoside at the 5'-position to the polymer support. P = Trityl functionalized polystyrene

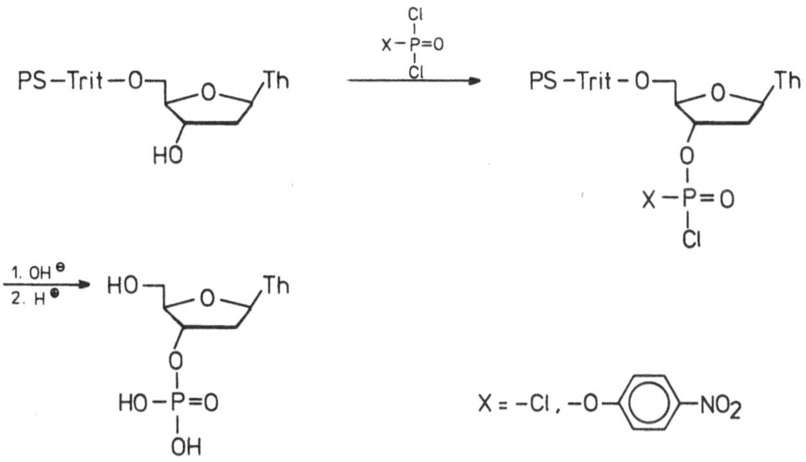

Scheme 7

4.3 Poly(vinyl alcohol) Carriers

In contrast to peptide synthesis, the application of insoluble supports on the basis of polystyrene has never found broad application [40, 59, 131, 132]. One of the main reasons for that is probably the steric hindrance by the structure of crosslinked polymers during condensation and cleavage reactions. Also, non-crosslinked, soluble carriers based on polystyrene are not especially appropriate because the solubility of loaded supports decreases with increasing numbers of condensation reactions. However, hydrophilic supports, such as poly(vinyl alcohol), have been used successfully for the synthesis of oligonucleotides.

Schott et al. [34, 48] showed that poly(vinyl alcohol) is suitable for the synthesis of short nucleotide blocks due to the high loading capacity of this support. They used commercially available poly(vinyl alcohol) ($\bar{M}$ = 63 000 g mol^{-1}), which was esterified with ribouridylic acid for the fixation of the first nucleotide, for the synthesis of thymidylphosphates with triisopropyl benzenesulfonic acid chloride as coupling reagent. The yields of the five condensation steps (12–36 h) were in the range of 10 to 30% as determined by the UV-absorption ratio. For long-chain oligonucleotides, however, polymeric supports with a lower capacity and solubility in pyridine are recommended [48].

Derivatives of poly(vinyl alcohol) were applied by Brandstetter et al. [152] as hydrophilic polymeric protecting groups. Reaction of 2-bromoethyl dihydrogenphosphate with PVA yielded carriers with up to 80% functionalization, whereas for other derivatives only 20% were attained after 24 h (Table 14). The PVA support ($\bar{M}$ = 70 000 g mol^{-1}) excels over the carriers based on poly(oxyethylene) in terms of the high loading capacity limited only by the solubilizing power of the polymer backbone.

Table 14. Derivatives of poly(vinyl alcohol) investigated in liquid-phase synthesis of oligonucleotides [152].

$$-(CH-CH_2)_n$$
$$|$$
$$O-R$$
$$\bar{n} = 1590$$

R	Yield in %
Br–CH$_2$–CH$_2$–	20
Br–CH$_2$CH$_2$–O–P with O double bond above and OH below	80
LiO–P(=O)(OLi)–SCH$_2$CH$_2$O–P(=O)(OH)–	20

Schott additionally demonstrated that one gram PVA can be loaded with up to 10 mmol of mononucleotide [34]. For this example he used as solvent mixtures of pyridine and hexamethylphosphorotriamide and noted condensation yields of more than 90%. Purification and removal of excess reagents was done by dialysis, precipitation or separation by column chromatography on Sephadex or DEAE-cellulose.

4.4 Poly(oxyethylene)-Based Supports

In addition to the soluble supports discussed in the preceding sections, hydrophilic polymers based on polyethers were introduced into the synthesis of oligonucleotides in order to overcome the problems of hydrophobic supports and carriers with a low solubilizing capacity [29, 31, 36].

Köster suggested poly(ethylene glycol) as macromolecular support for oligonucleotide synthesis [31]. He used a relatively high-molecular mass polyglycol ($\bar{M} = 20\,000$ g mol^{-1}) and introduced the monomethoxytrityl group *via* the tosylate and iodide derivative (Scheme 8). The loading capacity of 3'-O-acetyl thymidine was 28.7 μmol g^{-1}. Low-molecular weight compounds were eliminated by dialysis against a buffer solution in order to avoid damaging the labile oligonucleotides. Without any further functionalization reaction of the basis polymer he used the hydroxyl group directly with uridine as anchor [31]. The initial loading was 78.2 μmol g^{-1} and after coupling 127.5 μmol nucleotidic material was bound to 1 g support. This capacity determination served for the calculation of > 98% conversion of the condensation step (Scheme 9). According to the relatively low functional capacity of higher-molecular mass poly(ethylene glycol)s ($\bar{M} = 10\,000$ g mol^{-1}), which were used as diamino derivatives by Brandstetter et al. [28, 29], full solubility of the carrier-bound oligonucleotide throughout the synthesis was obtained. The overall yield of the trinucleotide was 14%:

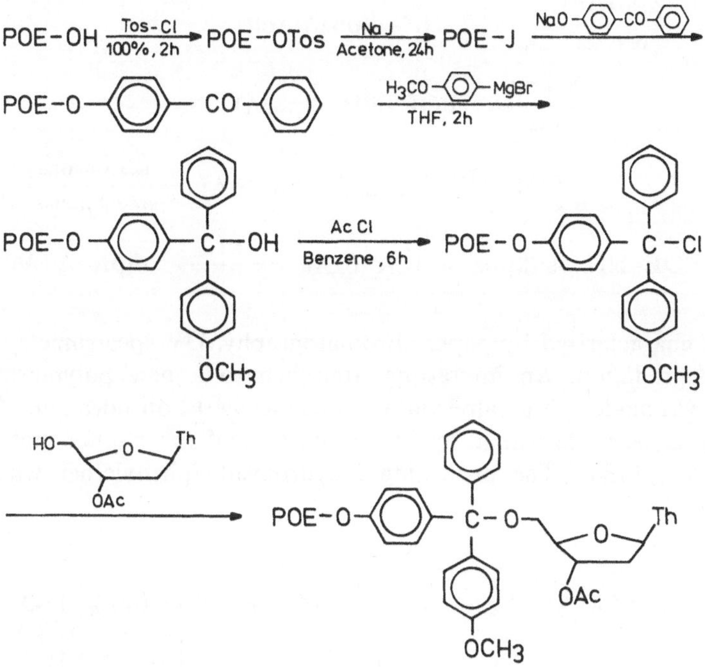

POE—OH $\xrightarrow[\text{100\%, 2h}]{\text{Tos-Cl}}$ POE—OTos $\xrightarrow[\text{Acetone, 24h}]{\text{Na J}}$ POE—J $\xrightarrow{\text{NaO}-\bigcirc-\text{CO}-\bigcirc}$

POE—O—$\bigcirc$—CO—$\bigcirc$ $\xrightarrow[\text{THF, 2h}]{\text{H}_3\text{CO}-\bigcirc-\text{MgBr}}$

POE—O—$\bigcirc$—C—OH $\xrightarrow[\text{Benzene, 6h}]{\text{Ac Cl}}$ POE—O—$\bigcirc$—C—Cl

OCH$_3$ OCH$_3$

$\xrightarrow{}$ POE—O—$\bigcirc$—C—O

OCH$_3$

Scheme 8

POE—OH $\xrightarrow[\text{TPS / Pyridine, 20h}]{}$ POE—O—P—O

$\xrightarrow[\text{1h}]{\text{0,1 N NaOH}}$ POE—O—P—O $\xrightarrow[\text{TPS, 4h, > 98\%}]{}$

POE—O—P—O

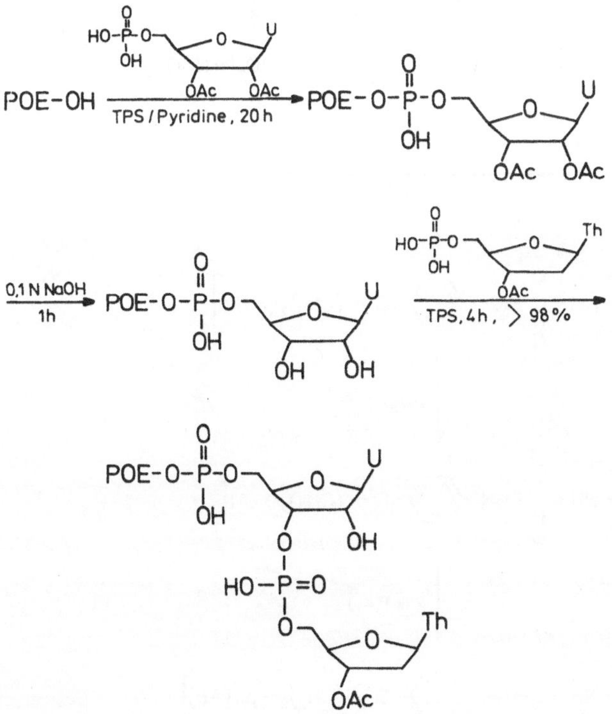

Scheme 9

$$\text{POE–NH}_2 \xrightarrow[\text{DCC, } 2d]{\text{d(pbzA(AC))}} \text{POE–NH–d(pbzA(Ac))}$$

$$\xrightarrow[\text{5 min}]{\text{1 N NaOH}} \text{POE–NH–d(pbzA)} \longrightarrow \longrightarrow$$

$$\text{POE–NH–d(pbz–A–bzA–bzA)} \xrightarrow[\text{HAc, Pyridine}]{\text{Isoamylnitrite}}$$

$$\text{POE–NH}_2 + \text{d(pbz–A–bzA–bzA)} \xrightarrow{\text{NH}_3} \text{d(pA–A–A)}$$

The product was characterized by paper chromatography, UV spectrometry, and enzymatic degradation. An interesting approach to a new polymeric protecting group was made by the same authors when using the diisocyanate of poly(oxyethylene) as basis derivative for the protection of the 5'-phosphate group of nucleotides [134]. The protecting 2-hydroxyethylphenylether was obtained as follows:

One of the main advantages of this polymeric protecting group is its cleavability under conditions where the protecting groups of nucleobases are stable. The cleavage was carried out with N-chlorosuccinimide and sodium hydroxide:

$$POE \left[NH - \underset{O}{\overset{\parallel}{C}} - NH - \bigcirc - S - (CH_2)_2 - pT(Ac) \right]_2$$

$$\downarrow$$

$$POE \left[NH - \underset{O}{\overset{\parallel}{C}} - NH - \bigcirc - SO_2 - (CH_2)_2 - pT(Ac) \right]_2$$

$$\downarrow$$

$$POE \left(NH - \underset{O}{\overset{\parallel}{C}} - NH - \bigcirc - SO_2 - CH = CH_2 \right)_2 + 2\ dpT$$

A very similar 2-hydroxythioether was described by Gait and Sheppard two years later, however, in combination with a highly-swellable, crosslinked poly(dimethyl acrylamide) [153]. Along the same line, the 2-hydroxyethylphenylether of poly(oxyethylene) was used as solubilizing polymeric protecting group for the synthesis of oligonucleotides [134].

Also, macromolecular protecting groups in the form of the thiophosphoric acid esters of POE were used for the liquid-phase synthesis of oligonucleotides [135, 152]. This derivative and also the 2-hydroxyethylphenyl thioether of POE are suitable as phosphate protecting groups because they react almost quantitatively with nucleotides. The introduction and cleavage of the polymeric groups is exemplified using deoxythymidine and 5'-deoxythymidylic acid. It was shown that it is possible to cleave the phosphate protecting group without affecting the protecting groups of the nucleo-base residues. The corresponding thiophosphates were obtained either by esterification of terminal nucleosides with hydrogen-S-alkylthiophosphates or by alkylation of 5'-thionucleotides with alkylhalides. Here, polymeric halides were allowed to react with trilithium monothiophosphate (Li_3PSO_3) yielding polymeric S-alkylthiophosphates. The yields were between 20 and 75%. A number of disubstituted poly(oxyethylene) derivatives, which have been applied in the synthesis of oligonucleotides, are listed in Table 15.

The conversions for the introduction of the polymeric phosphate protecting group into 3'-O-acetyldeoxythymidine and 3'-O-acetyldeoxythymidylic acid were in the range of 70 to 90%. Finally, 90% of the protecting groups could be cleaved with iodine after 6 h at room temperature [152].

Table 15. Bifunctional poly(oxyethylene) derivatives investigated in liquid-phase synthesis of oligonucleotides [152].

$$R-(CH_2CH_2O)_n-CH_2CH_2-R$$

$$\bar{n} = 135, 450$$

Basis derivatives R	Yield in %

$H_3C-\underset{\underset{O}{\|}}{C}-NH-\!\!\langle\bigcirc\rangle\!\!-O-$ −

$H_3C-\!\!\langle\bigcirc\rangle\!\!-SO_2O-$ 34

$Br-CH_2CH_2O-\underset{\underset{OH}{|}}{\overset{\overset{O}{\|}}{P}}-O-$ 90–100

$Cl-CH_2CH_2O-CONH-(CH_2)_6-NHCO-O-$ 88
$OCN-(CH_2)_6-NHCO-O-$ −

$LiO-\underset{\underset{OLi}{|}}{\overset{\overset{O}{\|}}{P}}-S-$ 30–40

$LiO-\underset{\underset{OLi}{|}}{\overset{\overset{O}{\|}}{P}}-S-CH_2CH_2O-\underset{\underset{OH}{|}}{\overset{\overset{O}{\|}}{P}}-O-$ 75

$\langle\bigcirc\rangle\!\!-S-CH_2\underset{\underset{CH_2OH}{|}}{CH}-O-$ 90

$\langle\bigcirc\rangle\!\!-S-CH_2\underset{\underset{OH}{|}}{CH}CH_2-O-$ −

$HO-CH_2CH_2-S-\!\!\langle\bigcirc\rangle\!\!-O-$ −

$HO-CH_2CH_2-S-\!\!\langle\bigcirc\rangle\!\!-NH-\underset{\underset{O}{\|}}{C}-NH-$ −

$-(CH_2)_6-NHCO-O-$ −

4.5 Miscellaneous Soluble Supports

There are a few soluble polymer supports which have only been reported once in the literature. However, in spite of this fact, some of them might be worth investigating in detail. Those discussed in this section are supports of the copolymer-type and poly(amino acid)s. Seliger and Aumann prepared a hydrophilic, non-crosslinked carrier based on poly(vinylpyrrolidone), which is soluble in pyridine and water, by bulk copolymerization of vinylpyrrolidone and vinylacetate with subsequent saponification of the acetate groups [36]. The functional capacity of hydroxyl groups was between 1.0 and 1.36 mmol g^{-1} [36]. The same type of copolymer was described before for peptide synthesis [15]. This is one of the very few polymer supports, which can be prepared directly by copolymerization (Scheme 10).

An interesting proposal was made for the use of a polypeptide support for the synthesis of a trinucleotide. Chapman and Kleid functionalized poly-L-lysine ($\bar{M} = 80\,000$ g mol^{-1}) in dimethylformamide solution to introduce an aromatic phosphoramidate linkage between the carrier and the nucleotide [154]. The polymer was crosslinked with hexamethylene diisocyanate yielding a product swellable in dimethylformamide and pyridine. Because at that time only solid-phase techniques were known, it has not yet been thoroughly investigated although this type of polar polymer support might be of potential benefit in a modified form for liquid-phase synthesis.

5. Outlook

The preparation and availability of suitable soluble polymer supports are basic requirements for the successful application of liquid-phase synthesis of peptides and nucleotides. Many parameters have to be considered when designing appropriate supports and the route of synthesis and the target product of synthesis must also be considered. Liquid-phase synthesis requires optimum solubility properties of the polymer supports and therefore the adaption of the functional capacity to the solubilizing power of the polymer backbone and side-chains is the determining step before synthesis. The main factors of evaluation are the solubility characteristics of the polypeptide or oligonucleotide to be synthesized.

So far, relatively few efforts have been made in terms of fundamental research on carrier problems of soluble polymers when compared to solid-phase synthesis. That could be a challenge for interdisciplinary research of both peptide or nucleotide chemists and polymer chemists.

Although a number of initial problems posed by the application of soluble polymer supports have been solved to a certain extent, the majority of difficulties in liquid-phase repetitive-type synthesis are still based on problems of carriers. Therefore, a greater variety of improved or new tailor-made soluble polymer carriers would help us to recognize the advantages of liquid-phase synthesis.

Generally, if there are no obstacles from the methodological aspects, striking improvements can still be made with regard to soluble supports, particularly

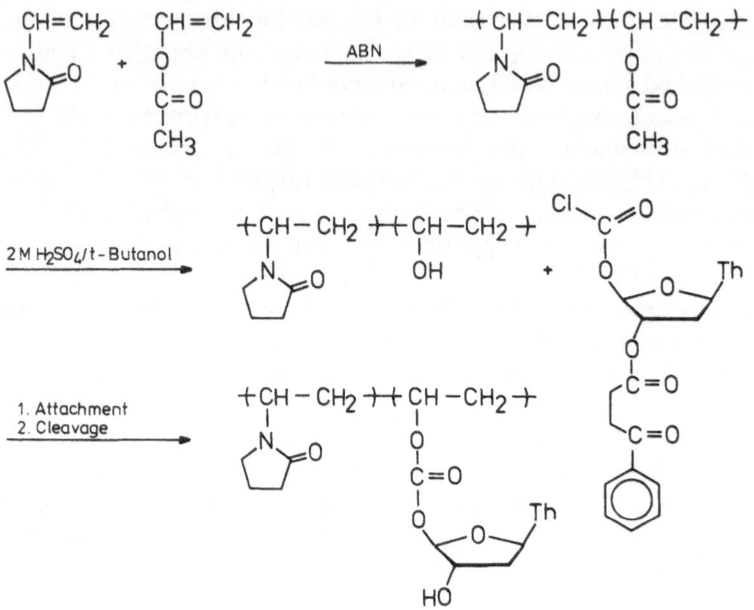

Scheme 10

when considering that the chemistry and applications of soluble polymers in general have only just gone through their infancy. From this point of view, further intense research and appropriate improvements of various parameters are needed to free the synthesis in the homogeneous phase at least from support-based problems. In addition to the increasing interest in the application of polymer supports, this might open access to other fields of support-based applications of soluble polymers.

6 References

1. Merrifield RB (1963) J Am Chem Soc 85: 2149
2. Letsinger RL, Kornet MJ (1963) J Am Chem Soc 85: 3045
3. Merrifield RB (1965) Science 150: 178
4. Stewart JM, Young GD (1969) Solid phase peptide synthesis, Freeman, San Francisco
5. Marglin A, Merrifield RB (1970) Ann Rev Biochem, 39: 841
6. Heitz W, Adv Polym Sci 23: 1 (1977)
7. Birr C (1978) Aspects of the Merrifield peptide synthesis, Springer, Berlin Heidelberg New York
8. Stahl GL, Walter R, Smith CW (1979) J Am Chem Soc 101: 5383
9. Smith CW, Stahl GL, Walter R (1979) Int J Pept Protein Res 13: 109
10. Geckeler K, Pillai VNR, Mutter M (1981) Adv Polym Sci 39: 65
11. Ovchinnikov YuA, Kiryushkin AA, and Kozhevnikova IV (1968) J Gen Chem USSR 38: 2551
12. Guthrie RD, Jenkin AD, Stehlicek J (1971) J Chem Soc (C) 2690

13. Mutter M, Hagenmaier H, Bayer E (1971) Angew Chem Int Ed 10: 811
14. Bayer E, Mutter M (1972) Nature 237: 512
15. Bayer E, Geckeler K (1974) Liebigs Ann Chem 1671
16. Mutter M (1978) Tetrahedron Lett 31: 2839
17. Geckeler K, Lange G, Eberhardt H, Bayer E (1980) Pure Appl Chem 52: 1883
18. Bayer E, Eberhardt H, Geckeler K (1981) Angew Makromol Chem 97: 217
19. Rivas, BL, Geckeler, KE (1992) Adv Polym Sci 102: 171
20. Bayer E, Schumann W, Geckeler K (1986) Chemically modified carbohydrates as highly efficient regio- and stereoselective catalysts. In: Carraher C, Sperling L (eds) Renewable resource materials. Plenum, New York, 115
21. Flory PJ (1939) J Am Chem Soc 61: 3334
22. Bayer E, Mutter M, Uhmann R, Polster J, Mauser H (1974) J Am Chem Soc 96: 7333
23. Pillai VNR, Mutter M (1982) Topics Curr Chem 106: 119
24. Andreatta RH, Rink H (1973) Helv Chim Acta 56: 1205
25. Mutter M (1972) PhD Thesis, University of Tübingen
26. Schattenberg M (1975) PhD Thesis, University of Tübingen
27. Mutter M, Uhmann R, Bayer E (1975) Liebigs Ann Chem 901
28. Brandstetter F (1972) Versuche zur Synthese von Oligonucleotiden an einem löslichen polymeren Träger, Diploma Thesis, University of Tuebingen
29. Brandstetter F, Schott H, Bayer E (1973) Tetrahedron Lett 32: 2997
30. Mutter M (1978) Tetrahedron Lett 31: 2843, (1978)
31. Köster H (1972) Tetrahedron Lett 16: 1535
32. Geckeler K (1975) (unpublished results)
33. Royer GP, Anantharamaiah GM (1979) J Am Chem Soc 101: 3394
34. Schott H (1973) Angew Chem Int Ed 12: 246
35. Geckeler K, Bayer E (1974) Prepr Int Symp on Macromolecules, Rio de Janeiro 244
36. Seliger H, Aumann G (1975) Makromol Chem 176: 609
37. Blecher H, and Pfaender P (1973) Liebigs Ann Chem 1973: 1263
38. Pfaender P, Pratzel H, Blecher H, Gorka G, Hansen G (1975) Pept Proc Eur Pept Symp 13: 137
39. Narita M (1978) Bull Chem Soc Japan 51: 1477
40. Cramer F, Helbig R, Hettler H, Scheit KH, Seliger H (1966) Angew Chem 78: 640
41. Hayatsu H, Khorana HG (1967) J Am Chem Soc 89: 3880
42. Katalog, Polyglykole Hoechst, Eigenschaften und Anwendungsgebiete der Polyethylenglykole, Farbwerke Hoechst, Frankfurt, 1969
43. Determann H (1970) Arch Pharm 303: 117
44. Strathmann H (1970) Chem Ing Tech 42: 1095
45. Bayer E, Mutter M (1974) Chem Ber 107: 1344
46. Geckeler K (1975) Dialog (Amicon) 10: 1
47. Geckeler K, Weingärtner K, Bayer E (1979) Prepr Int Symp on Polymeric Amines Gent 131
48. Schott H, Brandstetter F, Bayer E (1973) Makromol Chem 173: 247
49. Mutter M, Bayer E (1974) Angew Chem Int Ed. 13: 88
50. Hoffmann M, Kroemer H, Kuhn R (1977) Polymeranalytik I + II, Georg Thieme, Stuttgart
51. Bayer E, Gatfield J, Mutter H, Mutter M (1978) Tetrahedron 34: 1829
52. Bayer E, Zheng H, Albert K, Geckeler K (1983) Derivatives Polym Bull 10: 231
53. Bellof D, Mutter M (1984) Polym Bull 11: 49
54. Schmidt K, Geckeler K (1974) Anal Chim Acta 71: 79
55. Geckeler K (1979) Polym Bull 1: 427
56. Geckeler K (1976) unpublished results
57. Green B, Garson LR, J Chem Soc (C) 401: 1969
58. Geckeler K, Bayer E (1980) Polym Bull 3: 347
59. Hayatsu H, Khorana HG (1966) J Am Chem Soc 88: 3182
60. Köster H, Heyns K (1972) Tetrahedron Lett 1531
61. Geckeler K, Zheng H, Bayer E (1983) J Polym Sci Polym Chem Ed 21: 3541
62. Bayer E, Eckstein H, Hägele K, König WA, Brüning W, Hagenmaier H, Parr W (1970) J Am Chem Soc 92: 1735
63. Bayer E, Hagenmaier H, Jung G, Parr W, Eckstein H, Hunziker P, Sievers RE (1971) Proc Tenth Europ Pep Symp Abano Therme, Italy, Scoffone E, (ed.) North-Holland Publishing Company, Amsterdam, Netherlands 65
64. Atherton E, Clive DL, Sheppard RC (1975) J Am Chem Soc 97: 6584

65. Sparrow JT (1975) Chem Struct Biol Proc Am Pep Symp 4: 419
66. Varadarajan K, Fiat D (1983) Biopolymers 22: 839
67. Frank H, Hagenmaier H (1974) Tetrahedron 30: 2523
68. Hodge P (1979) Chem Ind (London) 18: 624
69. Mutter M (1979) Int J Pept Protein Res 13: 274
70. Bayer E, Mutter M, Holzer G (1975) Chem Struct Biol, Proc Am Pept Symp 4: 425
71. Schoknecht W (1979) Untersuchungen über die Anwendung der Liquid-phase-Peptidsynthese, PhD Thesis, University of Tuebingen
72. Schoknecht W, Albert K, Jung G, Bayer E (1982) Liebigs Ann Chem 1514
73. Geckeler K, Bayer E (1974) Makromol Chem 175: 1995
74. Kuenzi H (1978) Polymer compounds, Swiss patent 602 822
75. Anzinger H, Mutter M, Bayer E (1979) Angew Chem 91: 747
76. Anzinger H, Mutter M (1982) Polym Bull 6: 595
77. Colombo R (1981) Tetrahedron Lett 22: 4129
78. Pillai VNR, Mutter M, Bayer E, Gatfield I (1980) J Org Chem 45: 5364
79. Tjoeng F-S, Staines W, St-Pierre S, Hodges RS (1977) Biochim Biophys Acta 490: 489
80. Tjoeng F-S, Tong EK, Hodges RS (1978) J Org Chem 43: 4190
81. Tjoeng, F-S, Heavner GA (1982) Tetrahedron Lett 23: 4439
82. Colombo R, Pinelli A (1981) Hoppe-Seyler's Z. Physiol Chem 362: 1385
83. Ciuffarin E, Isola M, Leoni P (1981) J Org Chem 46: 3064
84. Bückmann AF, Morr M, Johansson G, Makromol Chem 182: 1379
85. Harris JM, Hundley NH, Shannon TG, Struck E (1982) J Org Chem 47: 4789
86. Geckeler K, Zheng H, Bayer E (1982) unpublished results
87. Bayer E, Zheng H, Geckeler K (1982) Polym Bull 8: 585
88. Geckeler K, Bayer E (1979) Polym Bull 1: 691
89. Bayer E, Grathwohl P, Geckeler K (1983) Makromol Chem 184: 969
90. Pillai VNR (1980) Synthesis, 1
91. Colombo R (1981) Hoppe-Seyler's Z. Physiol Chem 362: 1393
92. Glass JD, Silver L, Sondheimer J, Pande CS, Coderre J (1979) Biopolymers 18: 383
93. Heusel G, Jung G (1979) Liebigs Ann Chem 1173
94. Shemyakin MM, Ovchinnikov YuA, Kiryushkin AA, Kozhevnikova IV (1965) Tetrahedron Lett 27: 2323
95. Maher JJ, Furey ME, Greenberg LJ (1971) Tetrahedron Lett 1: 27
96. Maher JJ, Furey ME, Greenberg LJ (1971) Tetrahedron Lett 1: 29
97. Narita M, Hirata M, Kusano K, Itsuno S, Ue M, Okawara M (1980) Pept Chem 17: 107
98. Narita M (1979) Bull Chem Soc Jap 52: 1229
99. Narita M, Itsuno S; Hirata M, Kusano K (1980) Bull Chem Soc Jap 53: 1028
100. Isokawa S, Arai H, Narita M (1980) Pept Chem 17: 103
101. Sheppard RC (1983) Chem in Britain 5: 402
102. Isokawa S, Kobayashi N, Nagano R, Narita M (1984) Makromol Chem 185: 2065
103. Pfaender P, Pratzel H, Blecher H (1975) French Patent 2 243 183
104. Geckeler K, Bayer E (1974) Chem Ber 107: 1271
105. Frank H, Hagenmaier H (1975) Experentia 31: 131
106. Frank H, Meyer H, Hagenmaier H (1977) Chem Ztg 101: 188
107. Hagenmaier H, Frank H, Meyer H (1975) unpublished results
108. Jung G, Bovermann G, Göhring W, Heusel G (1975) Proc Fourth Am Pep Symp, Ann Arbor Science Publ Ann Arbor, 433
109. Heusel G, Bovermann G, Göhring W, Jung G (1977) Angew Chem Int Ed 16: 642
110. Heusel G (1978) Synthese von Peptiden durch Kombination von Polymerreagentien mit solubilisierenden Trägern, sowie der 2-Nitro-1-phenylethyl-Schutzgruppe, PhD Thesis, University of Tuebingen
111. Candau F, Afchar-Taromi F, Rempp P (1977) Polymer 18: 1253
112. Regen SL, Dulak L (1977) J Am Chem Soc 99: 623
113. Regen SL, Besse JJ, McLick J (1979) J Am Chem Soc 101: 116
114. Tsuchida E, Nishide H, Shimidzu N, Yamada A, Kaneko M, Kurimura Y (1981) Makromol Chem Rapid Commun 2: 621
115. Hefferman JG, Mackenzie WM, Sherrington DC (1981) J Chem Soc Perkin II Transact 514
116. Kimura Y, Regen SL (1983) J Org Chem 48: 195
117. Becker H, Lucas H.-W, Maul J, Pillai VNR, Anzinger H, Mutter M (1982) Makromol Chem Rapid Commun, 3: 217

118. Mutter M, Becker H, Lucas, H.-W, Maul J, Pillai VNR, Anzinger H (1982) Int Symp Polym Supp Reagents in Org Chem, Lyon, France 76
119. Cramer F (1966) Angew Chem 78: 186
120. Kössel H, Moon MW, Khorana HG (1967) J Am Chem Soc 89: 2148
121. Agarwal KL, Yamazaki A, Cashion PJ, Khorana HG (1972) Angew Chem 84: 489
122. Letsinger RL, Mahadevan V (1965) J Am Chem Soc 87: 3526
123. Letsinger RL, Mahadevan V (1966) J Am Chem Soc 88: 5319
124. Letsinger RL, Caruthers MH, Jerina DM (1967) Biochemistry 6: 1379
125. Shimidzu T, Letsinger RL (1968) J Org Chem 33: 708
126. Cramer F, Köster H (1968) Angew Chem 80: 488 (1968) Angew Chem Int Ed 7: 473
127. Melby LR, Strobach DR (1969) J Org Chem 34: 421
128. Freist W, Cramer F (1970) Angew Chem 82: 358
129. Sommer H, Cramer F (1972) Angew Chem 84: 710; (1972) Angew Chem Int Ed 11: 717
130. Glaser R, Sequin U, Tamm C (1973) Helv Chim Acta 56: 654
131. Kössel H, Seliger H (1975) Fortschr Chem Org Naturstoffe 32: 297
132. Amarnath V, Broom AD (1977) Chem Rev 77: 183
133. Ohtsuka E, Yamane A, Doi T, Ikehara M (1984) Tetrahedron 40: 47
134. Brandstetter F, Schott H, Bayer E (1974) Tetrahedron Lett 31: 2705
135. Brandstetter F (1974) Polymere Phosphatschutzgruppen zur Synthese von Oligonucleotiden in löslicher Phase, PhD Thesis, University of Tuebingen
136. Seliger H, Aumann G (1973) Tetrahedron Lett 31: 2911
137. Seliger H (1973) Makromol Chem 169: 83
138. Kabachnik MM, Polyakova IA, Potapov VK, Shabarova ZA, Prokof'ev MA (1970) Dokl Akad Nauk SSSR 195: 1344
139. Kabachnik MM, Potapov VK, Shabarova ZA, Prokof'ev MA (1971) Dokl Akad Nauk SSSR 201: 858
140. Potapov VK, Turkin SI, Shabarova ZA (1972) Zh Obshch Khim 42: 2349
141. Heidmann W, Koester H (1980) Makromol Chem 181: 2495
142. Itoh H, Ike Y, Ikuta S, Itakura K (1982) Nucleic Acid Res 10: 1755
143. Ohtsuka E, Morioka S, Ikehara M (1972) J Am Chem Soc 94: 3229
144. Köster H (1972) Tetrahedron Lett 16: 1527
145. Köster H, Geussenhainer S (1972) Angew Chem 84: 712; (1972) Angew Chem Int Ed 11: 713
146. Melby LR, Strobach DR (1967) J Am Chem Soc 89: 450
147. Melby LR, Strobach DR (1969) J Org Chem 34: 421
148. Melby LR, Strobach DR (1969) J Org Chem 34: 426
149. Köster H, Biernat J, McManus J, Wolter A, Stumpe A, Narang Ch, Sinha ND (1984) Tetrahedron 40: 103
150. Yip KF, Tsou KC (1971) J Am Chem Soc 93: 3272
151. Tsou KC, Yip KF (1973) J Macromol Sci A7: 1097
152. Brandstetter F, Schott H, Bayer E (1975) Makromol Chem 176: 2163
153. Gait MJ, Sheppard RC (1976) J Am Chem Soc 98: 8514
154. Chapman TM, Kleid DG (1973) J Chem Soc Chem Commun 193

Editor: Prof. H. Höcker
Received July 15, 1991

Polymers Containing Disulfide, Tetrasulfide, Diselenide and Ditelluride Linkages in the Main Chain

K. Kishore and K. Ganesh
Department of Inorganic and Physical Chemistry, Indian Institute of Science,
India, Bangalore 560012

The present review articulates the syntheses and properties of industrially important disulfide and tetrasulfide polymers. The diselenide and ditelluride polymers have also been reviewed, for the first time, so that a comprehensive view on the polymers containing group VIA elements can be obtained. The latter two polymers are gaining considerable current attention due to their semi-conducting properties. The emphasis has been made to sift through the developments in the last ten years or so to get the latest flavour in these rapidly developing polymers. We have also attempted to bring to the fore several contradicting results, like, for example, the crystallinity of ditelluride polymers, to clear the mist in such reports. We hope that this review will help those working in the field to assess the progress achieved in this area and that it may also provide useful orientation for those who wish to become involved.

1 Introduction

The main objective of the present review is to articulate the literature on the polymers of sulfur, selenium and tellurium covering mainly the disulfides, diselenides and ditellurides. The reason for restricting ourselves to only these polymers is to take a comprehensive view of the synthesis, stability and properties of these polymers containing the group VIA linkages. Although this type of group study is very common in simple organic compounds, it is rare in polymers. This laboratory has been working on the peroxide-containing polymers of group VIA [1–8] and a few review articles have also been published [9–12]. In order to get a comparative picture on the group of VIA polymers in its totality, it is necessary to have a systematic presentation of the literature on polymers containing disulfide, diselenide and ditelluride linkages at one platform.

The disulfide and tetrasulfide polymers are important industrially and they are used as sealants, adhesives, etc. The diselenide and ditelluride polymers are potentially attractive materials due to their semiconducting properties. In spite of these important features of the group VIA polymers, no complete review is available even on the polysulfides which have been known for a very long time. Most of the reviews on polysulfides are old [13–19] and a few later reviews [20, 21] do not provide a complete coverage of the latest developments. While sifting through the literature on the polysulfides, we have attempted to highlight important developments during the last ten years. On polyselenides and polytellurides, practically no reviews are available although some early information on the polymers containing selenium linkages is available in a book [22]. Several new methods have been found recently for preparing polyselenides and polytellurides and we have attempted to include these developments in the present review. In brief, this review gives a complete account of the synthesis and various studies carried out on polymers containing disulfide, tetrasulfide, diselenide and ditelluride linkages.

2 Polysulfides

2.1 Polysulfides from Sodium Polysulfide and Organic Dihalides

2.1.1 Historical Developments

The most widely used method for the preparation of the polysulfide polymers has been the reaction of inorganic polysulfide, usually sodium salts, with organic dihalides

$$Cl-R-Cl \ + \ Na_2S_x \ \longrightarrow \ (R-S_x)_n \ + \ 2\,NaCl \qquad (1)$$

where 'x' is referred to as the 'rank' and represents the average number of sulfur atoms in the polysulfide unit. The polymer derives its sulfide linkages from the alkali polysulfide, Na_2S_x, where sulfur rank varies from 2 to 4. The origin of polysulfide polymers produced by this method dates back to 1924–27 by Patrick and Mnookin [23], who carried out extensive studies on the reaction of organic dihalides with inorganic polysulfides. The first synthetic polysulfide rubber was marketed by the Thiokol company (U.S.A) in 1929 and they still remain the main producer even today. In fact the name 'Thiokol' has become synonymous to this class of polymer. The discovery of the synthetic rubber Thiokol A, i.e., poly(ethylene tetra sulfide) at the time when natural rubber was the only one available was of great significance. Thiokol A, because of its cheapness, was used in almost every application but soon it showed many disadvantages over the natural rubber, such as difficulty in processing, inferior physical properties and a disagreeable odour. Subsequently, significant improvement was made on the substitution of dichloroethane by bis(2-chloroethyl)ether, the tetrasulfide polymer derived from it was called Thiokol B and the disulfide polymer Thiokol D. The present commercial products are of the poly(ethyl formal disulfide) type.

The most significant improvement came in the early 1940s when a method for preparing thiol-terminated liquid polysulfides (Thiokol LP) was developed. The liquid polysulfides are particularly useful because the rubber could be compounded without using heavy mixing equipment. Cured thiokol LPs have high resistance to the environmental degradation, good low-temperature properties, low water-vapor transmission, good adhesion to wood, metal, glass and concrete and have excellent resistance to solvents, water, acids and bases [21].

2.1.2 Mechanism of Polymerization

Because the polymerization reaction (Eq. 1) involves inorganic salts of sodium polysulfide and organic dihalides, the literature has consistently treated polysulfide polymers as condensation polymers. Two features which show strong difference between polysulfides and typical polycondensation polymers are the high molecular weight obtainable in the polysulfide polymers and the effect of excess reactant on the molecular weight. Condensation polymers are rarely made with molecular weights of more than 30 000 because it is extremely difficult to attain stoichiometric proportion of the two reacting functional groups for various reasons, like, for example, purity and precise proportion of the reactants, occurrence of side reactions, etc. Polysulfide polymers on the other hand can be readily prepared with very high molecular weights than those achieved in a typical condensation polymerization. For example a low molecular weight polysulfide polymer was prepared and its molecular weight found to be 50 000 from osmotic pressure measurements in benzene solution [13]. It is reasonable to assume that the molecular weights of the high polymers are certainly over 500 000.

An excess of organic dihalide would produce a polymer of low molecular weight containing halide terminals. When the dihalide and sodium polysulfide are reacted in equimolar amounts, only low molecular weight ($\simeq 5000$) polymers are obtained [21]. The reason is as follows. Sodium polysulfide may be considered as a sodium salt of hydrogen polysulfide, which is a dibasic acid being slightly more acidic than hydrogen sulfide. Therefore the sodium salt undergoes extensive hydrolysis in aqueous solutions

$$Na_2S_x \; + \; H_2O \longrightarrow NaS_xH \; + \; NaOH \tag{2}$$

The relatively high concentration of OH$^-$ ions thus produced react competitively with the halides giving non-reactive terminal OH groups

$$-RCl \; + \; OH^- \longrightarrow -ROH \; + \; Cl^- \tag{3}$$

An excess of sodium polysulfide would produce a polymer of low molecular weight with terminal mercaptide groups (–SNa or –SSNa)

$$-RCl \; + \; Na_2S_2 \longrightarrow -RSSNa \tag{4}$$

These mercaptide terminals are oxidized by the excess of sodium polysulfide to produce the high molecular weight polymer

$$-RS_xNa \; + \cdot \; NaS_xR- \longrightarrow -RS_xR- \; + \; Na_2S_x \tag{5}$$

It has been established that polysulfide polymers in the presence of sodium polysulfide are constantly rearranging reversibly:

$$-RS_xR- \; + \; Na_2S_x \rightleftharpoons 2-RS_xNa \tag{6}$$

When the cleavage (Eq. 6) occurs near the end of a OH terminated oligomer (Eq. 7), the low molecular weight fragments are solubilized in the mother liquor. This removes the OH terminated fragments and allows the mercaptide-terminated fragments to react and continuously build the molecular weight. This phenomenon is referred to as toughening. The evidence for this comes from the experiment where polysulfide polymers containing deliberately placed OH terminals, on treatment with sodium polysulfide, quickly increase the molecular weight of the polymers [13]:

$$-RSSROH \; + \; Na_2S_2 \longrightarrow -RSSNa \; + \; NaSSROH \tag{7}$$
$$\text{(into aq. phase)}$$

After washing the sodium polysulfide and the solubilized OH–terminated fragments, the molecular weight of the resultant polymer is considered to be at least 500 000–1 000 000 and the yield is about 80%. The above studies clearly demonstrate as to why high molecular weight polymer is obtained and assert that the role of sodium polysulfide is not merely a reactant like the organic dihalide. In the other methods of preparing polysulfide polymers, where sodium polysulfide is not used, the increase in molecular weight by solubilization of terminals does not occur.

2.1.3 Choice of Monomers

In spite of the variety of difunctional reagents such as tosylates, sulfates, thiosulfates and even gem-dinitrates, the ready availability and the reactivity of dihalides, especially the dichlorides, have made them the reagents of choice for preparing polysulfide polymers [20]. The relative reactivities of the halides with alkali polysulfides follow the pattern expected for nucleophilic substitution reactions. Alkyl bromides are more reactive than alkyl chlorides, and alkyl fluorides are generally unreactive [24]. Aromatic and vinyl halides, unless strongly activated by substituent groups [25], are also unreactive under normal reaction conditions. The order of reactivity for alkyl halides is primary > secondary > tertiary [24]. The secondary and tertiary halides are generally unsuitable as monomers because of their tendency to undergo dehydrohalogenation reactions [20]. With the exception of dichloro- and dibromo-methane, gem-dihalides do not give polymeric products under normal reaction conditions [13, 26]. Other than dihalides [27], polychlorohydrocarbons [28, 29] like pentachloronitrobenzene [30] have also been used to prepare the polymers. Monomers such as bis(2-chloroethyl)sulfide('mustard gas'), glyceroldichlorohydrin (the polymer is moisture sensitive due to high hydroxyl content) and chloroacetate derivatives of glycerol, etc., have been used to prepare polymers having specific properties [20, 31].

Although sodium sulfide is commonly used in inorganic polysulfide synthesis, other alkali ammonium and alkaline earth polysulfides can also be used provided they are sufficiently soluble [31]. Sodium polysulfide solutions are prepared by reacting aqueous NaOH solution with sulfur, their ratio determining the sulfur rank:

$$6\,NaOH \; + \; (2x + 1)\,S \;\longrightarrow\; 2\,Na_2S_x \; + \; Na_2SO_3 \; + \; 3\,H_2O \quad (8)$$

When excess sulfur is used, it reacts with sodium sulfite to give sodiumthiosulfate:

$$S \; + \; Na_2SO_3 \;\longrightarrow\; Na_2S_2O_3 \quad\quad\quad\quad\quad\quad\quad (9)$$

To retard the above reaction, sodium sulfite can be crystallized from the solution by using concentrated NaOH and carrying out the reaction at high temperatures [32]. The excess NaOH is then destroyed by reacting it with sodium hydrosulfide:

$$NaOH \; + \; NaSH \;\longrightarrow\; Na_2S \; + \; H_2O \quad\quad\quad\quad\quad (10)$$

Sodium polysulfide solutions can also be prepared by reacting sodium sulfide or sodium hydrosulfide with sulfur in water [20]:

$$Na_2S \; + \; (x - 1)\,S \;\longrightarrow\; Na_2S_x \quad\quad\quad\quad\quad\quad\quad (11)$$

$$2\,NaSH \; + \; (x - 1)\,S \;\longrightarrow\; Na_2S_x \; + \; H_2S \quad\quad\quad\quad (12)$$

2.1.4 Sulfur Rank

The aqueous solutions of sodium polysulfide, Na_2S_x, prepared according to the above procedures contain a distribution of sulfide anions ranging from S^{2-} to S_5^{2-} and possibly higher [17]:

$$S^{2-}, S_2^{2-}, S_3^{2-}, S_4^{2-}, S_5^{2-}, HS^-, HSS^-, HS_3^-, HS_4^- \text{ and } HS_5^-$$

Since a distribution of all these species always exists, the term "sulfur rank" merely represents the average number of anions present. Thus, when Na_2S_2 is used, one always obtains a distribution of sulfide linkages, a majority of which are disulfides [16]. This is also reflected in the polysulfide elastomers which have approximately the same sulfur rank as the sodium polysulfide. However, when excess of sodium polysulfide is used, the resulting polymer generally has a lower sulfur rank than the corresponding alkali polysulfide. This important phenomenon of polysulfide polymers is called 'desulfurization' [16]. A contradicting trend was found by Voronkov et al. [33] when they polymerized 9, 10-dibromo-anthracene with Na_2S_n (n = 2–5). The average degree of sulfidity of the copolymer increased by 25–90% with increase in the molar ratio of Na_2S_n: dibromide, at constant n. The polysulfide bridge was found to increase linearly with the initial value of n. Such increase in length was attributed to the radical processes occurring during polycondensation.

2.1.5 Desulfurization and Thionation

The sulfur in polysulfide polymers, even those of high rank, is chemically bound as evidenced by the failure to remove free sulfur by solvent extraction [16]. However, it is possible to remove sulfur from the polysulfide linkages by using chemical agents like alkali hydroxides, hydrosulfides and sulfites [18] in a process called "desulfurization" or "stripping". Alkali sulfite is by far the most preferred, economically as well as for ease of handling:

$$-R-S-S-S-R- \quad + \quad SO_3^{2-} \quad \longrightarrow \quad -R-S^- \quad + \quad {}^-O_3S-S-S-R$$

$$\longrightarrow \quad -R-S-S-R- \quad + \quad S_2O_3^{2-} \quad (13)$$

Similarly, NaOH [34] and Na_2S can also be used:

$$4\left(R-S_4\right)_n \quad + \quad 6n\,NaOH \quad \longrightarrow \quad \left(R-S_2\right)_n \quad + \quad n\,Na_2S_2O_3$$

$$+ \quad 2\,Na_2S_3 \quad + \quad 3n\,H_2O \quad (14)$$

$$\left(R-S_4\right)_n \quad + \quad n\,Na_2S \quad \longrightarrow \quad \left(R-S_2\right)_n \quad + \quad n\,Na_2S_3 \quad (15)$$

Sulfur can be stripped off from the polysulfide polymer up to the disulfide stage only [18]. The reverse of Eq. (15) represents a method for increasing the rank of the polymer by using an excess of a higher rank alkali polysulfide. This unique process is called "thionation". The increase in polymer rank by this technique is limited to a maximum of approximately 4 to 4.5 due to the instability of aqueous inorganic sulfides of rank higher than five [16]. Thionation can also be achieved by milling the polymer with sulfur in presence of a base catalyst like diphenyl guanidine [16]. Polymers containing monosulfide linkages do not undergo desulfurization or thionation reactions.

2.1.6 Reactivity of Sodium Polysulfide

Since an aqueous solution of inorganic polysulfide ions is an equilibrium mixture of monosulfide to pentasulfide ions, it is very difficult to study the relative reactivity of the individual polysulfide ions since they cannot be isolated to permit kinetic studies [19]. Regarding the reactivity, the nucleophilicity of the sulfide anions increases with the rank [20]. The equilibration is effected by the nucleophilic displacement interchange at the organic polysulfide bond:

$$-RSS \; SR- \;\; + \;\; \;^-S_y \;\; \longrightarrow \;\; -RS \; SS_y^- \;\; + \;\; \;^-SR$$

$$\longrightarrow \;\; -RSSR- \;\; + \;\; \;^-SS_y^- \qquad\qquad (16)$$

It was found that, irrespective of the course of condensation reaction, the interchange yields a polymer mixture with a rank corresponding to the equilibrium value with the ionic polysulfide environment [19].

2.1.7 Ring Formation

As in the other polycondensation reactions, competition between linear polymerization and ring formation also occurs here and it is strongly dependent on the structure of both the dihalide and inorganic polysulfide. Cyclic monosulfides are favoured when the dihalide monomer has four or five carbon atoms between the halide terminals [18].

2.1.8 Preparation of Polysulfide Polymers

The first commercial alkyl polysulfide polymer was prepared from the reaction of ethylene dichloride with aqueous sodium tetrasulfide. The use of methanol or ethanol solution in place of aqueous polysulfide did not prove to be advantageous [31]. To prevent the polysulfide polymer, which is insoluble in the medium, from separating in the form of lumps during the reaction, an emulsion of the condensates is made by adding emulsifying agents like magnesium hydroxide [35, 36], barium sulfate [31, 18], glue, gelatin, methyl cellulose, casein and dextrin [31]. Fine emulsions can be obtained with higher alkyl sulfonic acids and dibutylnaphthalene sulfonic acid [31]. A typical polymerization based on sodium tetrasulfide and sodium disulfide is reported elsewhere [35, 37].

Random copolymers can be prepared either by polymerizing two different halides with polysulfide [38, 39] or by redistributing the blends of homopolymer lattices in the desired molar proportions (Eq. 17). The latter procedure is preferred with the dihalide polymers since differential solubility of the monomer intermediates may result in the enrichment of a monomer component in the final polymer [16]:

$$-S-SRS-SRS-S- \ + \ -S-SR'S-SR'S-S- \ \longrightarrow \ -S-SR'S-SRS-SRS-S-$$

$$(17)$$

Another interesting method of preparing copolymers is reported by Todorova et al. [40–43] which involves the interaction of vinyl monomers with a polysulfide polymer. Under dispersion conditions breaking of polysulfide bonds occur forming ions, $RS_{1 \text{ to } 3}^{+}$ $RS_{3 \text{ to } 1}^{-}$ and radicals, $RS \cdot$, $RSS \cdot$, which then react with styrene or acrylonitrile [40] to form the copolymer:

$$-R-S_4-R- \ + \ H_2C = \underset{\underset{CN \ (or \ Ph)}{|}}{CH} \ \longrightarrow \ -R-S_n-CH_2-\underset{\underset{CN \ (or \ Ph)}{|}}{CH}-S_{4-n}-R- \qquad (18)$$

$$n = 1, 2, 3$$

Block copolymers are prepared by coupling two thiol terminated homo-polymers by oxidation [31]:

$$H(SSR)_n SH \ + \ HS(R'SS)_m H \ \xrightarrow{\text{(0)}} \ H(SSR)_n SS(R'SS)_m H \qquad (19)$$

2.1.9 Structure

2.1.9.1 End Groups

Generally, high molecular weight polysulfide polymers are insoluble in prac-tically all organic solvents which prevent their characterization by solution techniques. The end groups in a polysulfide polymer chain were originally thought to be either thiol groups, produced by the excess of sodium disulfide, or the unreacted chlorine groups. Later evidence showed that this was not true, as low molecular weight polymers, which contain no chlorine by analysis, could be appreciably increased in molecular weight by treating with additional sodium polysulfide [13]. The presence of thiol terminals also seemed unlikely, as the treatment with ordinary oxidizing agents such as H_2O_2 did not appreciably increase the molecular weight [13]. Other possibilities considered were un-saturated terminals produced by dehydrohalogenation of the terminal chlorine by the alkaline environment or hydroxyl groups produced by alkaline hydroly-sis of an organic halide. Hydroxy compound formation has been reported in some cases (Eq. 20) [44] and certain other studies also indicated that the polysulfide polymers contain hydroxy terminals [45]:

$$Cl \text{—}\langle A \rangle\text{—} Cl \xrightarrow{\text{Na}_2\text{S}_4} Cl\text{—}\langle A \rangle\text{—}S_4\text{—}\langle A \rangle\text{—}OH \qquad (20)$$
$$O_2N \quad NO_2 \qquad\qquad H_2N \quad NH_2 \; H_2N \quad NH_2$$

It was also found that diisocyanates could produce appreciable increases in molecular weight [13]. The action of sodium polysulfide in increasing molecular weight was not at all pronounced when a hydrophobic terminal, such as hydrocarbon, was placed at the end of the chain, whereas hydroxyl terminals could be removed by treating with sodium disulfide [13]. It is possible to prepare polysulfide polymers with other functional groups by treatment with either thiols or disulfides containing the desired functional group. For example, hydroxyl terminals can be readily placed at the end of a polymer chain by treating with a disulfide containing hydroxyl groups such as dithioglycol, which readily interchanges with the disulfide groups in the polymer to give a polymer containing essentially hydroxyl terminals [37, 46, 47].

2.1.9.2 Polysulfide Linkage

Early work on the tetrasulfide and disulfide polymers led to the conclusion that in tetrasulfide the two sulfurs were postulated as being linked to the side of the two sulfurs which are present in the backbone. Hence, the chain segment itself was a disulfide group (1). The chief reason for this belief was the easy removal of two of the sulfurs from a polysulfide polymer by treatment with sodium hydroxide or by a sulfur acceptor such as sodium sulfite [48].

$$-C-S-S-C- \qquad\qquad -C-S-S-S-S-C-$$
$$\quad\;\; \| \;\; \| $$
$$\quad\;\; S \;\; S$$

$$\textbf{1} \qquad\qquad\qquad\qquad \textbf{2}$$

Evidence from X-ray diffraction was interpreted to support this structure [49, 50]. Later X-ray evidence seemed to indicate that tri and tetrasulfides have almost a linear arrangement (2) [51, 52] and more studies on the polymers and simple organic polysulfides support this view [53–57].

2.1.9.3 Cross-Linking

The common technique which is used for making synthetic elastomers is to have unsaturation in the polymer backbone whose vulcanization is usually achieved by opening the double bond leading to a cross-linked product. This technique has not been found useful in vulcanizing polysulfide elastomers. While it is possible to put unsaturation in the polymer by using monomers like 1,4-dich-loro-2-butene [13], this has not resulted in the successful production of a polymer which can be cured by sulfur and accelerator combinations. In a few cases compounds containing actively polymerizable vinyl groups, like for example divinyl polysulfides, were prepared from vinyl epoxy ethers and sodium polysulfide (Eq. 21). The curing is then achieved by polymerizing through the double bonds [58]. Similarly, (vinylaryl)alkyl-terminated polysulfides have been prepared by reacting sodium polysulfides with vinylbenzyl chloride (Eq. 22) or with a mixture of vinylbenzyl chloride and other aliphatic dihalides [59]:

$$H_2C = CHOROCH \underset{O}{\overset{}{\diagdown\!\diagup}} \;\; + \;\; Na_2S_x \;\; \longrightarrow \;\; (H_2C = CHOROCHCH_2 \underset{OH}{}\!)_{\overline{1/2}} \, S_x \quad (21)$$

$$2\,H_2C = CH - C_6H_4CH_2Cl \;\; + \;\; Na_2S_x \;\; \longrightarrow \;\; (H_2C = CH - C_6H_4 - CH_2\!)_{\overline{1/2}} \, S_x$$

$$(22)$$

The cured vinyl terminated polymers are reported to have low odour and, when blended or copolymerized with other polymers, they impart many beneficial properties of polysulfide polymers such as resistance to oxygen permeation, water, UV light and solvents [60–63].

The original vulcanizing agent used was zinc oxide which appears to effect a coupling between the terminal hydroxyl groups [45]. Polymer containing pendant hydroxyl groups was prepared by using 1,2-dichloropropanol-3, and the resultant polymer was later reacted with difunctional acids [64] or anhydride [65] to produce a network structure. This method has not been found very practical in actual applications. Previous efforts to introduce network formation by adding polyhalides with functionality greater than two were unsuccessful because the high molecular weight products could not be processed [20]. However, with the advent of thiol terminated liquid polysulfide (LP) polymers, all the above-mentioned problems have nearly been sorted out.

2.1.10 Thiol Terminated Liquid Polysulfide (LP) Polymers

Cleavage of high molecular weight rubber to a low molecular weight liquid polysulfide (LP) polymer actually involves chemical reduction of some of the disulfide links to thiol terminals. The degree of this reduction determines the molecular weight of the resulting liquid polymers. The chemistry of this reduction depends on the interchange reaction between sodium hydrogen sulfide and the disulfide groups:

$$-SRS-SRS \;\; + \;\; NaSH \;\; \rightleftharpoons \;\; -SRSNa \;\; + \;\; HSSRS- \quad (23)$$

$$-SRSNa \;\; \overset{H^+}{\longrightarrow} \;\; -SRSH \; ;$$

Acid or bisulfite is generally added to neutralize the sodium mercaptide terminals formed [17]:

$$-SRSNa \;\; + \;\; NaHSO_3 \;\; \longrightarrow \;\; -SRSH \;\; + \;\; Na_2SO_3 \quad (24)$$

For each –SH terminal formed, a disulfide –SH terminal is also formed. Addition of a quantity of sodium sulfite ensures that the polysulfide –SH terminal is reduced to a simple thiol:

$$-RSSRSSH \quad + \quad Na_2SO_3 \quad \longrightarrow \quad -RSSRSH \quad + \quad Na_2S_2O_3 \qquad (25)$$

Besides, the sulfite also removes the polysulfide polymer ranks greater than 2. Other than sodium hydrogen sulfide–sodium sulfite mixture [66–70], sodium hydroxide–sodium dithionite [68, 69, 71], sodium sulfide–sodium sulfite [66, 72] and hydrazine-sodium hydroxide mixtures [73, 74] have also been tried as reducing agents. Typical procedures for the preparation of liquid polysulfide polymers have been reported elsewhere [31, 75, 76].

Liquid polysulfides can also be prepared directly with thiol end groups by reacting a mixture of aliphatic di- and polyhalides with a mixture of sodium disulfide and sodium hydrogen sulfide [77]. Reaction of polymeric disulfides with polymercaptans also leads to the formation of liquid polysulfides [37]. LP polymers with $\bar{M}_n$ in the range of 600 to 8000 are generally preferred. The uncured LP polymers are soluble in toluene, benzene and chlorinated hydrocarbons such as ethylene chloride [17, 78]. LP polymers like LP-31, LP-2, LP-32, LP-12, LP-3, LP-33, LP-5, LP-8, based on range of viscosities, molecular weights and functionalities, are available [17].

2.1.10.1 Viscosity-Molecular Weight Relationship

The polysulfide liquid polymers have a unique viscosity-molecular weight relationship; the viscosity varies approximately as the cube of the molecular weight [17]:

$$\log \mu \quad = \quad \log K \quad + \quad n \log \bar{M}_n \qquad (26)$$

where 'μ' is the viscosity expressed in $dP_{a.s}(= P)$ at 80 °C, the constant 'n' is 2.75 ± 0.03 and the constant 'K' is $5 \pm 1 \times 10^{-8}$. $\bar{M}_n$ is calculated from the thiol content of the polymer assuming the functionality of two. The relationship applies only to liquid polymers with the normal Gaussian distribution of molecular sizes.

2.1.10.2 Cure of LP Polymers

The most common way to convert LP polymers to solid elastomers is to couple the terminal –SH groups. This is done either by oxidation to disulfides using organic or inorganic oxidizing agents or by reaction with epoxy resins, aromatic amines, diisocyanates, aldehydes, etc [16, 35]:

$$2 -RSH \quad \xrightarrow{(O)} \quad -R-S-S-R- \quad + \quad H_2O \qquad (27)$$

$$\triangle_O \quad + \quad HS-R-SH \quad \longrightarrow \quad -C-C-S-R-S-C-C- \qquad (28)$$
$$\qquad\qquad\qquad\qquad\qquad\qquad\quad OH \qquad\qquad\qquad OH$$

Oxidizing agents include lead dioxide, manganese dioxide, tellurium dioxide, zinc peroxide, calcium peroxide, zinc chromate, alkali chromates, cumene hydroperoxide and organic tin compounds [18, 31]. Activators like NH_3, diphenylguanidine, quinone dioxime, DMF, water and 2,4,6-trisdimethylaminomethylphenol are also sometimes used [31].

Lead dioxide cures the polymer at room temperature in the presence of moisture in a weakly alkaline medium [16]:

$$2-RSH \quad + \quad PbO_2 \quad \longrightarrow \quad -RSSR- \quad + \quad PbO \quad + \quad H_2O \qquad (29)$$

Lead oxide formed may react in turn with thiols to give metallic thiolates which are detrimental [17]; addition of sulfur removes them as metal sulfides:

$$2-RSH \quad + \quad PbO \quad \longrightarrow \quad -RSPbSR- \quad + \quad H_2O \qquad (30)$$

$$-RSPbSR- \quad + \quad S \quad \longrightarrow \quad -RSSR- \quad + \quad PbS \qquad (31)$$

Curing is usually carried out in a formulation with fillers, plasticizers and curing-rate modifers like stearic acid [17].

2.1.11 Properties

2.1.11.1 Glass Transition Temperature and Crystallinity

The glass transition temperature is dependent on the nature of the hydrocarbon moiety and the polysulfide rank. The greater the hydrocarbon proportion the lower the T_g while higher the polysulfide rank the higher the T_g [16]. The melting is more difficult to identify since the crystallization rate of these polymers is usually very slow. Prolonged storage at low temperatures brings about partial crystallization. The monomer units containing relatively high proportions of the hydrocarbon and the most regular structures are more easily crystallized. For example, while poly(ethylene disulfide) crystallizes only to a small extent (14%), that too at a very low rate [79], the pentamethylene and hexamethylene disulfide polymers harden fairly rapidly at $-20\,°C$ [16]. Polymers based on dichloroethylether and dichloroethylformal, although flexible at low temperatures, have no tendency for crystallization [13]. Disulfide polymers of δ-chlorobutylether and δ-butylformal harden by crystallization in the range -20 to $-40\,°C$ [13]. In copolymers, while the random copolymer containing 75/25 mol ratio of poly(butyl formal disulfide) and poly(ethyl formal disulfide) did not show any crystallization, the corresponding block disulfide copolymer of the same composition showed a small degree of crystallization [16].

2.1.11.2 Stability of Polysulfide Polymers

The disulfide linkages in polysulfide polymers is susceptible to cleavage by hot concentrated alkali. Amines were claimed to be solvents for the polysulfide elastomers, but actually they cleave the disulfide bonds and lower their molecular weight [79]. The disulfide links are generally stable to acids, except for the acids which oxidize the disulfide linkage and cleave the chain. The present commercial polymers, which are based on bis(2-chloroethyl)formal, contain formal linkage which is more susceptible to hydrolysis by acids. Actually, hydrolysis of high molecular weight polymers has been used to prepare low molecular weight polymers with terminal hydroxyl groups [13].

Because of their saturated hydrocarbon structure and the antioxidant nature of sulfur, the polysulfide polymers are extremely resistant to aging and to oxidation [13]. Polymers which have a short hydrocarbon repeating units between sulfur atoms such as methylene, ethylene or trimethylene are not stable to high temperatures due to the formation of stable volatile cyclic materials. Introducing longer segments into the polymer chain greatly decreases the tendency for weight loss at high temperatures and the odour is believed to be due to the formation of cyclic materials [13]. The thermal stability of polymethylene and polyethylene polysulfides was found to decrease with increase in the sulfide linkage from 1 to 4 [80]. Polysulfide polymers having either ether or the formal linkages are less thermally stable compared to those containing saturated hydrocarbon segments [13]. Since the commercially available polymers are based on the poly(ethyl formal disulfide) backbone, extensive studies on its thermal stability has been done by Rosenthal and Berenbaum [16]. They reported that the initial attack on this structure is an acid catalyzed hydrolytic attack on the formal group by trace amount of water:

$$-SCH_2CH_2OCH_2OCH_2CH_2S- \; + \; H_2O \; \longrightarrow \; 2 -SCH_2CH_2OH \; + \; CH_2O$$

$$(32)$$

The reaction releases free formaldehyde which then reduces the backbone disulfide bonds to thiol with the formation of formic acid:

$$-CH_2CH_2SSCH_2CH_2- \; + \; CH_2O \; \longrightarrow \; 2 -CH_2CH_2SH \; + \; HCOOH \quad (33)$$

Formic acid further catalyzes the hydrolysis at the formal group. The mercaptan terminals react in turn with the hydroxyl group to form monosulfide linkages with the evolution of water which is then available for further hydrolysis:

$$-CH_2CH_2SH \; + \; HOCH_2CH_2S- \; \longrightarrow \; -CH_2CH_2SCH_2CH_2S- \; + \; H_2O$$

$$(34)$$

The above mechanism has been confirmed from model studies where disulfide, formaldehyde and water heated in a sealed tube produced a reduction of the disulfide to thiol and an oxidation of formaldehyde to formic acid [13].

The consequences of these processes is a weight loss together with hardening of the polymer due to the monosulfide structure. The acidity required to initiate the polymer hydrolysis arises from the peroxidation of the formal bond and subsequent breaking of this intermediate to formic acid and water [16]. This is a free-radical process and can be inhibited using an antioxidant such as phenyl-β-naphthylamine [13]. In practical systems there is no significant effect of the antioxidants due to the excess of oxidizing agents being used to bring about commercially useful cure rates. Instead, compounds such as diepoxides or diisocyanates, which will convert the mercaptan to high molecular weight polymer without attacking the formal or the disulfide bonds in the backbone shows substantially improved high temperature performance [16]. Another source of thermal instability arises from the presence of mercaptide groups which undergo degradation by cyclodepolymerization [16]:

$$-SSCH_2CH_2OCH_2OCH_2CH_2S \overset{\frown}{-} S \overset{\displaystyle CH_2CH_2O}{\underset{\displaystyle S-CH_2CH_2O}{\diagup}} CH_2$$

$$\longrightarrow -SSCH_2CH_2OCH_2OCH_2CH_2S^- \; + \; \overset{S}{\underset{S}{\mid}} \overset{\displaystyle CH_2CH_2O}{\underset{\displaystyle CH_2CH_2O}{\diagup}} CH_2 \qquad (35)$$

Since the mercaptide group is regenerated, the interchange can continue until a major part of the polymer is volatilized [16]. Pyrolysis of cured polysulfide polymers based on bis(2-chloroethyl)formal has also been studied under GC-MS conditions by Rama Rao et al. [81]. Apart from the cyclic monomer, appreciable amounts of other compounds are also formed. The breaking of formal bond was found to be the initiation step.

2.1.11.3 Viscoelastic Properties

The viscoelastic properties of the polysulfide polymers are governed by the interchange reactions of the polysulfide linkages in the chain. On rapid application of stress, the cross-linked rubbers behave as conventional elastomers. Under a fixed strain, the following interchange processes set in to relieve the stress:

$$RSSR \; + \; R'SSR' \; \rightleftharpoons \; 2\,RSSR' \qquad (36)$$

$$RSSR \; + \; R'SH \; \rightleftharpoons \; RSSR' \; + \; RSH \qquad (37)$$

The first reaction needs sodium disulfide for the interchange whereas for the second reaction the interchange takes place even in its absence [37]. Disulfide–disulfide interchange is relatively slow at temperatures below 150 °C or in the absence of UV light. Trace amounts of sulfur, mercaptides or alkaline agents capable of generating mercaptides by reaction with disulfide bonds catalyze the exchange reaction and bring about stress relaxation at moderate temperatures [16]. Since it is very difficult to eliminate the cure accelerator residues, they catalyze the exchange process.

The effect of curing agents on the relaxation behaviour of polysulfide polymers has been studied [82] on the following three cure systems: lead peroxide, manganese dioxide plus morpholine and 2,4-toluene diisocyanate plus N-methyl-2-pyrollidone. Despite the variation in the linkage formation rates, the first two cures showed the same activation energy of 24 kcal/mol which probably corresponds to the mercaptide-disulfide exchange process. The cure involving the condensation of the mercaptan terminals with a diisocyanate completely avoids the presence of the ionic structures or ionic impurities. In this case higher activation energy of 36.6 kcal/mole was obtained and attributed to a disulfide–disulfide interchange process.

Tobolsky et al. [83] have studied the stress relaxation of poly(ethylene tetrasulfide) cross-linked with 10% of trichloropropane. In the temperature range 92–104 °C, the tetrasulfide polymers relaxed at fifteen times the rate of the disulfide polymer. However, high molecular weight disulfide and tetrasulfide polymers showed the same activation energy, 24 kcal/mol, for the relaxation rate. Since the activation energy for a pure disulfide polymer exchange should be closer to 36.6 kcal/mol, it was concluded that the relaxation process in the high molecular weight disulfide polymers probably occurs by the exchange of tri and tetrasulfide bonds. Addition of 5 wt% of sulfur to poly(ethylene disulfide) reduces the relaxation time from 470 to 180 s at 100 °C. However, no effect was observed on the addition of sulfur to poly(ethylene tetrasulfide) [83].

2.1.11.4 Molecular Weight Distribution

The polysulfide linkages in high molecular weight polysulfide polymers due to both polysulfide–polysulfide and mercaptan-disulfide interchange reactions result in molecular weight redistribution. The term polysulfide used here refers to the sulfur linkages of rank two and higher. Using chromatographic fractionation techniques Genkin et al. [84] calculated the polydispersity coefficients ($\bar{M}_w/\bar{M}_n$) to be 1.06 to 1.55 for mercaptan terminated polysulfide polymers ($\bar{M}_n = 1170$–2010), indicating very narrow molecular weight distributions. Shlyakhter et al. [85] showed the fractional composition of polymers obtained by mixing samples of different viscosities and molecular weights to be similar to that of individual polymers of the same viscosity. It was therefore concluded that thiol-disulfide interchange occurs when the polymers are mixed, leading to a narrow molecular weight distribution. Conversely, the polydispersity coefficient of mercaptan terminated polysulfide liquid polymers ($\bar{M}_n = 3700$–6000) was found by Fettes and Mark [86] to be in the range of 2.8 to 6.2. However, Shlyakhter and coworkers [87] using the same procedure as Fettes and Mark [86] found the polydispersity to be approximately 2 indicating random molecular weight distribution. When the mercaptan terminals were converted to thioacetyl groups, the interchange reactions did not take place even at higher temperatures. It was concluded that mercaptan-disulfide interchange takes place very readily during the preparation of mercaptan terminated polymers as well as in the bulk.

2.1.11.5 Conformational Analyses and Photo Stability

Polarization infrared spectral data, X-ray analysis and normal coordinate treatment revealed the stable molecular conformation of poly(methylene disulfide) to be the GG'G form [88] which was also confirmed from the results of semi-empirical CNDO/2SCF MO calculations [89]. Poly(ethylene disulfide) was also found to exist in a similar conformation as that of poly(methylene disulfide) [90]. Under vacuum at 50 °C, polysulfide polymers of methylene and ethylene with sulfur rank of two and four were exposed to UV radiation [91]. While poly(methylene disulfide) and poly(methylene tetrasulfide) yielded polymeric carbon monosulfide, hydrogen sulfide and carbon disulfide as the major degradation products, the ethylene counterparts produced the same compounds except carbon disulfide. The tetrasulfide polymers also formed volatile products which on condensation gave the original polymer.

2.2 Polysulfides from Condensation of Poly(thiodiglycols)

The monomer used in this method is a difunctional compound which has a di- or polysulfide linkage like bis(2-hydroxyethyl)disulfide, commonly known as 'dithiodiglycol', which was traditionally made by the reaction of ethylenechlorohydrin with sodium disulfide [13]. Betrozzi [92] prepared this compound by the reaction of ethylene oxide and sodium sulfide:

$$2 \: H_2C - CH_2 \: + \: Na_2S_2 \: + \: 2 H_2O \longrightarrow HOCH_2CH_2SSCH_2CH_2OH \: + \: 2 NaOH$$
$$\underset{O}{\diagdown\diagup}$$

$$(38)$$

By this method, it proved impossible to prepare the polythiodiglycol with a rank higher than about 3.0, even by using sodium tetrasulfide [13]. Tetrathiodiglycol was prepared in non-aqueous medium as tetrathiodiglycollic acid by the reaction of sulfur monochloride with the disodium salt of thioglycolic acid [13]. When the tetrasulfide is added to water, two atoms of sulfur are liberated and the dithiodiglycolic acid is left as the highest sulfide capable of existing in an aqueous medium [13].

   The reaction of dithiodiglycol and formaldehyde produces a polymeric formal containing disulfide groups [93]:

$$HOCH_2CH_2SSCH_2CH_2OH \: + \: HCHO \longrightarrow -(CH_2CH_2OCH_2OCH_2CH_2SS)-$$

$$+ \: H_2O \quad (39)$$

which has a structure similar to the polymer formed from the reaction of bis(2-chloroethyl)formal with sodium disulfide:

$$ClCH_2CH_2OCH_2OCH_2CH_2Cl \quad + \quad Na_2S_2 \quad \longrightarrow \quad \{CH_2CH_2OCH_2OCH_2CH_2SS\}_{\overline{n}}$$

$$+ \quad 2\,NaCl \quad (40)$$

It is much easier to produce polymers of high molecular weight by the latter method than by the former because of the effect of excess sodium polysulfide.

Dithiodiglycol can be converted into a polymeric disulfide containing an ether linkage by utilizing the enhanced reactivity of a hydroxyl group on a carbon beta to the sulfur atom [94–97]:

$$HOCH_2CH_2SSCH_2CH_2OH \quad \longrightarrow \quad \{CH_2CH_2OCH_2CH_2SS\}_{\overline{n}} \quad + \quad H_2O \quad (41)$$

This unusual easy cleavage from a carbon atom of a primary hydroxyl group was first found by Bennett and Hock [98, 99] in the self condensation of 2-mercaptoethanol:

$$HOCH_2CH_2SH \quad \longrightarrow \quad \{CH_2CH_2S\}_{\overline{n}} \quad + \quad H_2O \quad (42)$$

Ethylene or diethylene glycol was also used to effect the copolymerization with di- or polythiodiglycol [100–102]. Salts of di- or polythiodiglycol were also used to prepare polymers by condensing with organic dihalides like 1,2-dichloroethane, bis(2-chloroethyl)ether, or bis(2-chloroethyl)sulfide [103, 104]. The reaction of the metallic salt of a glycol with bis(2-chloroethyl)sulfide [103, 105] was also used to prepare similar polymers. Andrews et al. [106, 107] studied the acid-catalyzed polymerization of dithioglycol and found that phosphorus pentoxide efficiently catalyzes the etherification reaction [106]. He also prepared copolymers with thiodiglycol and dithiodiglycol.

2.3 Polysulfides from di-Bunte Salts

Early work [108–110] showed that polymeric disulfides could be prepared from sodium polysulfide and organic dithiosulfates or "di-Bunte salt":

$$NaO_3S{-}SRS{-}SO_3Na \quad + \quad Na_2S \quad \longrightarrow \quad \{SRS\}_{\overline{n}} \quad + \quad n\,Na_2S_2O_3 \quad + \quad n\,Na_2SO_3$$

$$(43)$$

Di Bunte salt can be readily made from the reaction of sodium thiosulfate and organic dichlorides or dibromides [111]:

$$Cl{-}R{-}Cl \quad + \quad 2\,Na_2S_2O_3 \quad \longrightarrow \quad NaO_3S{-}SRS{-}SO_3Na \quad + \quad 2\,NaCl \quad (44)$$

This method is particularly useful in making the polymers which otherwise cannot be made by the sodium polysulfide process. For example, action of sodium polysulfide on 1,4-dichlorobutane produces only the cyclic sulfide and thiacyclopentane [13]. Using sodium thiosulfate first to form the dithiosulfate, the polymer was subsequently obtained by treatment with sodium disulfide [13]. This method is also advantageous in that less hydrolysis of the halide function occurs compared to the reaction between a dihalide and sodium disulfide. However, this method may also give cyclic sulfide if R is short enough [21]. From the IR spectra the thiosulfuric group was found to be the most plausible end group in these polymers [112]:

$$-RSSO_3^- \xrightarrow{H^+} -RSSO_3H \tag{45}$$

But even by this method pure disulfide polymer cannot be obtained, due to the entering of the sulfur atom of sodium sulfide nanohydrate into the polymer chains [113] thus producing a polymer containing polysulfide linkages. It was also observed that the sulfur content in the polymer increases with increase in the concentration of sodium sulfide used [112]. Instead, when dimethylsulfoxide was used, pure disulfide polymer could be obtained [112]. Copolymers of polyxylylene and polyethylene disulfide have also been prepared using this method [114].

2.4 Polysulfides from Cyclic Disulfides

An extremely interesting method for the preparation of polymeric disulfides is the self-polymerization of cyclic disulfides [115–117]:

$$(H_2C)_2 \overset{O}{\underset{S-S}{\diamond}} (CH_2)_2 \longrightarrow \{SCH_2CH_2OCH_2CH_2S\}_n \tag{46}$$

It is however much more difficult to prepare the cyclic monomer. The general procedures for preparing cyclic disulfides involves: depolymerizing high molecular weight polymers [115–118], treatment of organic thiosulfate with cupric salts followed by steam distillation [119], reaction of the dibromide with an alcoholic solution of sodium disulfide [120] and oxidation of dithiols [121, 122]. Polymeric species are also formed in the oxidative reaction [123]. Hence precautions must be taken to prevent polymerization during the synthesis.

2.4.1 Stability of Cyclic Disulfides Towards Polymerization

The stability of a cyclic disulfide is dependent on its structure and purity. Lipoic acid is relatively stable in the solid state [120]. If, however, the disulfide is heated above its melting point, polymerization readily occurs [123]. For the disulfides of C_5–C_{12} n-alkyl α,ω-dithiols, both the six- and seven-membered ring compounds are reported to be stable in pure forms [124]. Conversely, another study on the same series of disulfides reports that only seven-, nine- and eleven-membered rings are stable in the pure form [119]. Another study reports only

3,3,5,5-tetramethyl-1,2-dithiocylcopentane and 1,2-dithiane to be stable to polymerization [125]. These conflicting reports probably reflect the differences in the purities of the disulfides.

Literature on the stability of individual cyclic disulfides covers: (i) compounds that polymerize on standing like 3,3-dimethyl-1,2-dithiolane [124, 125], 3-hydroxy-1,2-dithiolane [126] and N-phenyl-4-oxo-5-aza-1,2-dithiacyclohexane [127, 128]; ii) compounds which polymerizes on heating such as dibenzo-1,2,5,6-tetrathiocin [129] and iii) compounds that polymerize in neat form when a catalyst is added, such as cis-1,2-dithiacyclohex-4-ene [130] and 1-oxa-4,5-dithiacycloheptane [13, 119, 130]. Detailed study of the stability of cyclic disulfides towards polymerization revealed that 1,2-dithianes are the only class of simple cyclic disulfides that is thermodynamically stable with respect to its polymerization in the melt or as concentrated solutions [131].

Cyclic disulfides have been converted to polymeric disulfides by ionic catalysts such as H_2SO_4, $FeCl_3$ (with a trace of water), $AlCl_3$, BF_3, and $SnCl_4$ [132]. Other than Lewis acids, various alkaline catalysts were also used as initiators for the polymerization [133]. Active compounds are, for example, alkali thiolates, sulfides, alcoholates, hydroxides and various amines. Dainton et al. [134] determined the heat of polymerization of several cyclic disulfides. The six-membered ring showed the lowest heat of polymerization. The ability of cyclic disulfides to copolymerize free radically with common monomers [135] such as styrene [136] and vinyl acetate [137] has been demonstrated. Copolymers have also been prepared by terminating the living vinyl polymers with polydisulfides [138] as well as by using polydisulfides as radical chain transfer agents for the growing vinyl polymer chain [139].

2.4.2 Flourinated Polydisulfides

Refluxing a reservoir of sulfur (445 °C) and passing the tetrafluoroethylene directly into the hot sulfur vapor gives three monomers namely tetrafluoro-1,2,3,4-tetrathiane (cyclic tetrasulfide), tetrafluoro-1,2,3-trithiolane (cyclic trisulfide) and tetrafluorothirane (cyclic sulfide) respectively [140]:

$$S_8 \;+\; F_2C=CF_2 \;\longrightarrow\; \cdot S_x CF_2 CF_2 \cdot \tag{47}$$

$$\cdot S_x CF_2 CF_2 \cdot \;\longrightarrow\; \underset{F}{\overset{F}{\diagup}}\!\!\diagdown \text{(tetrathiane)} \;+\; \text{(trithiolane)} \;+\; \text{(thirane)} \tag{48}$$

Tetrafluorotetrathiane polymerizes in acetonitrile at $-40\,°C$ to form highly crystalline poly(tetrafluroethylene tetrasulfide):

$$\text{(tetrafluorotetrathiane)} \;\xrightarrow[-40°C]{CH_3CN}\; \{CF_2CF_2S_4\}_n \tag{49}$$

Synthesis of the polymeric disulfide was achieved by heating an equimolar mixture of cyclic tetrasulfide with tetrafluoroethylene at 200 °C. The cyclic tetrasulfide probably undergoes homolysis at the isolated or 2,3 sulfur–sulfur bond giving relatively stable thiyl radicals which then interact with tetrafluoroethylene, possibly forming an eight-membered ring intermediate, to yield the polymer:

$$\text{(cyclic tetrasulfide)} \xrightleftharpoons{F_2C=CF_2} \left[\text{(eight-membered ring intermediate)}\right] \longrightarrow \left(SCF_2CF_2S\right)_n \qquad (50)$$

Thermal reaction of the cyclic trisulfide with tetrafluoroethylene yields 1,2,5-trithiepane. This compound polymerizes with triethylamine at $-40\,°C$ to form a disulfide polymer which also contains a monosulfide linkage:

$$\text{(cyclic trisulfide)} + F_2C=CF_2 \longrightarrow \text{(ring)}$$

$$\longrightarrow \left(CF_2CF_2SCF_2CF_2S_2\right)_n \qquad (51)$$

The disulfide polymer is also formed from the pyrolysis of the monosulfide polymer which was prepared from the ring opening polymerization of tetrafluorothiirane by radical means:

$$\text{(tetrafluorothiirane)} \xrightarrow{h\nu} \left(CF_2CF_2S\right)_n \xrightarrow{\Delta} \left(CF_2CF_2S_2\right)_n$$

$$+ \quad F_2C=CF_2 \quad + \quad \text{(ring)} \qquad (52)$$

Monomeric cyclic tetrasulfide distills off along with other compounds and sulfur from the tetrasulfide polymer at 250 °C. But when pyrolyzed under pressure at 300 °C, octafluoro-1,4-dithaine and sulfur are formed:

$$\left(CF_2CF_2S_4\right)_n \xrightarrow[\text{distill}]{200-250\,°C} \text{(ring)} + \text{ etc.}$$

$$\xrightarrow[8h]{300\,°C} \text{(ring)} + S_8 \qquad (53)$$

Under vacuum at 250 °C the disulfide polymer degrades to form a dimer of tetrafluoro-1,2-dithietane and dithietane. When the pyrolysis is conducted in the presence of hexafluoropropylene, a 1,4-dithiane bearing a trifluoromethyl group is obtained:

$$(54)$$

2.5 Polysulfides based on Condensation Type Monomers Containing a Disulfide Group

2.5.1 Polyurea and Polyquinazolone

Fedotova et al. [141] reported the synthesis of polyurea from the reaction of hexamethylenediisocyanate and bis(4-aminophenyl)disulfide:

$$(55)$$

The disulfide polyurea has a molecular weight of 8000, is soluble in H_2SO_4 and cresol and starts decomposing at 220 °C. The 4,4'-diaminodiphenyl-3,3'-dicarboxylic acid and 4,4'-diacetamidodiphenyldisulfide react to give dark amber glass-like prepolymer [142]:

$$(56)$$

which tautomerizes to **3**

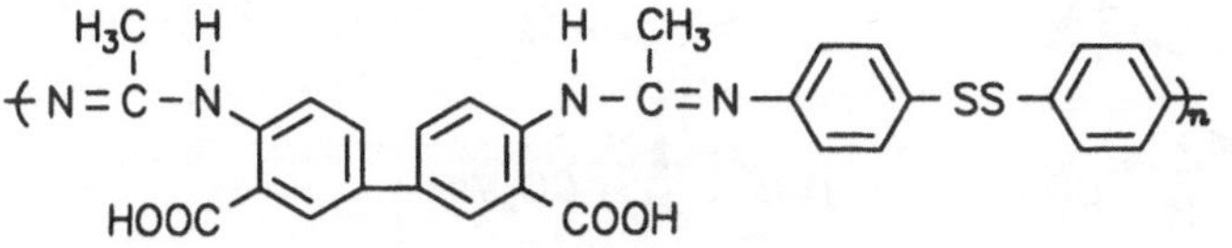

H₃C H   H CH₃

+N=C-N...phenyl rings...N-C=N...phenyl rings...SS...phenyl ring +ₘ

HOOC   COOH

3

The presence of both forms have been established by IR spectroscopy. Cyclo-dehydration of the prepolymer yielded polyquinazolone (**4**).

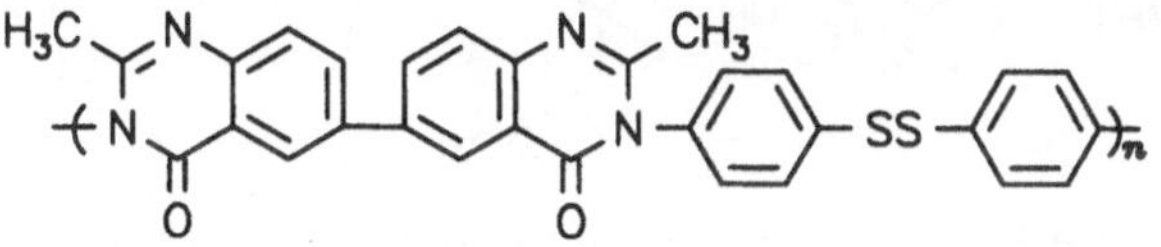

H₃C...N...   N...CH₃

+N...   N...phenyl...SS...phenyl +ₘ

O   O

4

Due to the presence of flexible disulfide linkage, the cured polymer showed increased ductility and decreased hardness.

2.5.2 Polyamides and Polyimides

Synthesis of polyamides containing disulfide linkages were achieved by condens-ing 2,2′-dithiobis(acetyl chloride) and 3,3′-dithiobis(propionyl chloride) with diamines like hexamethylene diamine [143]. The polymers are crystalline and can be drawn into fibers. The disulfide bonds in the main chain are dissociated by mercaptan exchange reactions [144]:

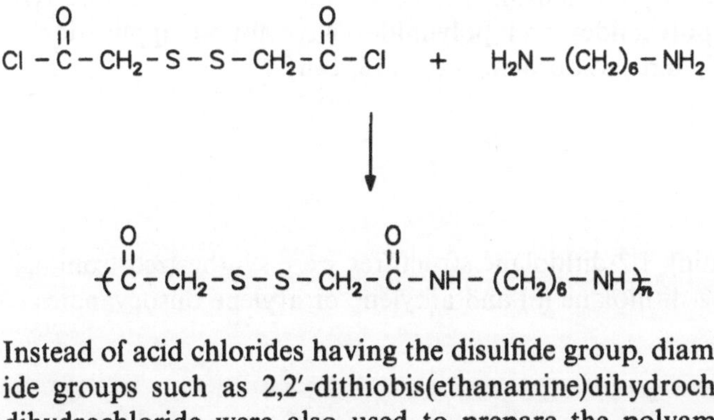

$$Cl-\overset{O}{\overset{\|}{C}}-CH_2-S-S-CH_2-\overset{O}{\overset{\|}{C}}-Cl \quad + \quad H_2N-(CH_2)_6-NH_2$$

$$\downarrow$$

(57)

$$+\overset{O}{\overset{\|}{C}}-CH_2-S-S-CH_2-\overset{O}{\overset{\|}{C}}-NH-(CH_2)_6-NH+_n$$

Instead of acid chlorides having the disulfide group, diamines containing disulf-ide groups such as 2,2′-dithiobis(ethanamine)dihydrochloride and cystamine dihydrochloride were also used to prepare the polyamides [145, 146]. Both diamine as well as the diacid possessing the disulfide group such as bis-D-penicillamine methyl ester and N,N′-bisphenylacetyl-L-cystine were also condensed to produce the polyamide [147]:

$$(58)$$

Harris and Eury [148] prepared biodegradable disulfide polymers by the ring opening polymerization of oxazolones with diamines:

$$(59)$$

The polyimide is made by condensing the acid chloride with aqueous NH_3 [143]. Some of the polyamides and polyimides have found application as antistatic agents [149] and radioprotectants [145, 146].

2.5.3 Polyurethanes

Polyurethanes containing 1,2-dithiolane structures were synthesized from 4,4'-bis(hydroxymethyl)-1,2-dithiolane (5) and alkylene or arylene diisocyanates.

Cross-linked polyurethanes have also been prepared by adding a triol. Reduction of the polymers was carried out by using NaH in DMSO or $LiAlH_4$ in THF. The reduced polyurethanes were used as substitutes for lipoic acid derivatives in the reduction of protein thiol bridges [150].

2.5.4 Poly(Schiff Bases)

Reaction of terephthaladehyde with 4,4'-diaminodiphenyldisulfide forms a Schiff base type polymer [151]:

OHC—⟨benzene⟩—CHO + H_2N—⟨benzene⟩—S–S—⟨benzene⟩—NH_2

$$\downarrow \tag{60}$$

⟨CH—⟨benzene⟩—CH=N—⟨benzene⟩—S–S—⟨benzene⟩—N⟩ₙ

This polymer along with other added compounds, increases the resistance towards scorch and thermal aging in rubbers. Similarly, condensation of 2,5-dihydroxyterephthalaldehyde with 2,2'-diaminodiphenydisulfide also yields a Schiff base type polymer, which forms a black nickel chelate with Ni salts [152].

2.6 Polydisulfides from Dithiols

Although the oxidation of thiols to disulfides is well known, there are only a few instances where this method has been used for making polysulfide polymers [153–157]. Marvel and Olson [158] found that aerial oxygen was a better reagent for oxidizing the dithiols to polymeric disulfides than Br_2, HNO_3 or $FeCl_3$. The polymers obtained were of low molecular weights:

$$HS-(CH_2)_6-SH \xrightarrow{(O)} H\{S-(CH_2)_6-S\}_n H \tag{61}$$

Oxidation of dithiols by air to polydisulfides in the presence of a mixture of tertiary amine and a copper salt [159, 160] or in the presence of an inorganic basic compound has also been achieved [161]. Unlike bromine, iodine was found to have efficiently oxidized the dithiols to polydisulfides [162, 163]. Reduction of disulfide polymers back to dithiols was achieved either by using Na/liquid NH_3 [164, 165] or with a pyridine/acetic acid/zinc mixture [166]. When DMSO was used as the reaction medium, it also acted as an oxidizing agent of dithiols to polydisulfides [167–169]. For example, poly(m-phenylene disulfide) prepared by the oxidation of 1,3-dimercaptobenzene with DMSO

[170] as such is an insulator, but, when complexed with chlorosulfonic acid, its conductivity increases by 11 orders of magnitude. Poly(m-phenylene disulfide), in presence of sulfur and $AlCl_3$ at 80 °C in n-heptane, forms intramolecularly cyclized poly(phenylene sulfide) of the thianthrene-type [171]. Other than dithiols of hydrocarbons, even dithiols containing heteroatoms such as 1,3-di(2-mercaptoethyl)tetramethyl disiloxane and 1,3-di(mercaptopropyl)tetramethyl disiloxane were also oxidized to yield polydisulfides containing siloxane units [172]. The main difficulty in this method of polymerization is the preparation of pure thiols and the oxidation must proceed without any side reactions so that high molecular weights are achieved.

2.6.1 Oxidative Coupling of Dithiols with Diacyl Disulfides

Bis(oxycarbonyl)disulfide reacts with dithiols in the presence of triethylamine to give disulfide polymers [173]:

$$RO-\overset{\overset{O}{\|}}{C}-S-S-\overset{\overset{O}{\|}}{C}-OR \quad + \quad HS-R'-SH$$

$$\longrightarrow \quad \{R'-S-S\}_n + \quad 2\,COS \quad + \quad 2\,ROH \tag{62}$$

This method is called fragmentation polymerization since each step of the chain extension involves thiol-induced heterolytic fragmentation of $-SSC(=O)OR$ moieties:

$$X-\overset{\overset{X}{\|}}{C}-S-S-\overset{\overset{Y}{\|}}{C}-X \quad + \quad HS-R'- \quad \xrightarrow[-XH]{-CYS} \quad -R'-S-S-\overset{\overset{Y}{\|}}{C}-X \tag{63}$$

$$-R'-S-S-\overset{\overset{Y}{\|}}{C}-X \quad + \quad HS-R'- \quad \xrightarrow[-XH]{-CYS} \quad -R'-S-S-R'- \tag{64}$$

Other than bis(oxycarbonyl) disulfides, compounds such as thiuram disulfides, dixanthogens and bis(carbomyl) disulfides were also used to prepare disulfide polymers [174]. Alternating as well as random copolymers have also been prepared using this method [175, 176].

2.7 Polysulfides from Sulfur Monochloride

Poly(xylene tetrasulfide) has been made by condensing p-xylenedithiol with sulfur monochloride, S_2Cl_2, in CS_2 solution [177]:

$$HS-CH_2-\!\!\left\langle\!\!\bigcirc\!\!\right\rangle\!\!-CH_2-SH \quad + \quad S_2Cl_2 \quad \longrightarrow \quad \{CH_2-\!\!\left\langle\!\!\bigcirc\!\!\right\rangle\!\!-CH_2-S_4\}_n$$

$$\tag{65}$$

Highly crystalline poly(p-phenoxydisulfide) was prepared by the reaction of diphenyl ether and S_2Cl_2 in chloroform in the presence of traces of iron catalyst [178]. Reaction of phenols with S_2Cl_2 in an inert solvent yields disulfide polymers [179]. Similarly by reacting alkyl phenols with S_2Cl_2 in the presence of basic catalysts, polyhydroxylic and alkyl-substituted products were obtained [180, 181]. Using alkali metal hydrocarbon complexes of diisodiooctadiene, disodiophenylbutane or sodium napthalene and S_2Cl_2, the resulting polymers can be used as solid rocket-fuel binders, oil and solvent-resistant gaskets and sealing compounds [182].

Poly(*tert*-butylamine sulfide) prepared from *tert*-butylamine and S_2Cl_2 and poly(imine sulfide) prepared from amylamine and S_2Cl_2 were found to be superior vulcanizing agents for rubbers compared to sulfur and other poly(alkylamine sulfides) [183, 184]. Paushkin et al. [185–193] have done extensive studies on poly(amine sulfides). Cross-linked polyethylene, on heating at 150 °C for 75 h in air in the presence of poly(aniline disulfide), lost only 5% of the crosslinked phase in comparison to 100% in the absence of poly(aniline disulfide) [185]:

$$\langle \hspace{-0.3em} \bigcirc \hspace{-0.3em} \rangle - NH_2 \;+\; S_2Cl_2 \;\longrightarrow\; \{ \langle \hspace{-0.3em} \bigcirc \hspace{-0.3em} \rangle - NH - S - S \}_n \tag{66}$$

Stabilization of the thermal degradation of polyethylene by poly(p-phenylenediamine disulfide) and poly(iminodisulfide) has also been studied [185, 186]. The mechanism is the formation of weakly active S radicals, consequently leading to slowing down of the reaction. Poly(arylamine disulfides) were also found to have efficiently catalyzed the dehydration of diacetone alcohol [187, 188] and 1-phenyl cyclohexanol [189, 190] respectively. Since poly(arylamine disulfides) are paramagnetic, their electrical and magnetic properties such as electrical conductivity (σ) and concentration of ferromagnetic species (N) respectively were also investigated [191, 192]. The value of σ at 30 °C ranged from 2.9×10^{-6} to 3.6×10^{-9} $\omega^{-1} cm^{-1}$ for bis(p-aminophenyl)methane-sulfur-monochloride copolymer and p-aminoazobenzene-sulfurmonochloride copolymer respectively. The values of N were nearly the same (ca. 10^{19} spin/g) for polymers containing only NHSS bridges between the aromatic rings, e.g., for poly(aniline disulfide). The N value ranged from 5.9×10^{16} to 1×10^{20} spin/g for polymers containing additional bridges (O, SO_2, CH_2, NH or NN) between the aromatic rings as in bis(p-aminophenyl)sulfone-sulfur monochloride copolymer (**6**).

$$\{ HN - \langle \hspace{-0.3em} \bigcirc \hspace{-0.3em} \rangle - SO_2 - \langle \hspace{-0.3em} \bigcirc \hspace{-0.3em} \rangle - NH - S - S \}_n$$

6

Since paramagnetic semiconducting polymers, having conjugated chains, catalyze the dehydration of 2-methyl-3-buten-2-ol to isoprene and *tert*-BuOH to isobutylene, the effect of inorganic oxides which have paramagnetic or diamagnetic phase was studied [193]. While the ceramic superconductor

Bi–Sr–Ca–Fe–CuO_x, having a paramagnetic phase, exhibited catalytic activity in the dehydration of *tert*-BuOH to isobutylene, another ceramic superconductor Bi–Sr–Co–Fe–CuO_x, having a diamagnetic phase, inhibited the reaction. The paramagnetic poly(aniline disulfide), when used as a modifier for LaHY Zeolite (20%)-Al_2O_3 (80%) catalysts for the cyclohexane isomerization at 350 °C, increased the selectivity for methylcyclopentane and methylcyclopentenes [194].

Polydisulfides containing hydroxyl groups have been synthesized by condensing 3,3′-diamino-4,4′-dihydroxydiphenysulfone:

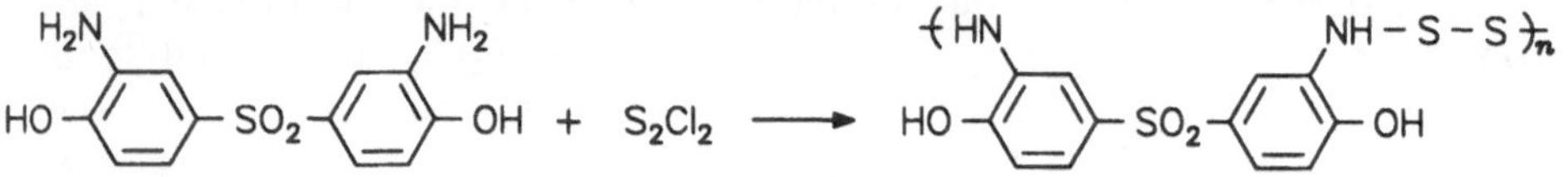

(67)

or 4,4′-dihydroxydiphenylsulfone with S_2Cl_2 [195].

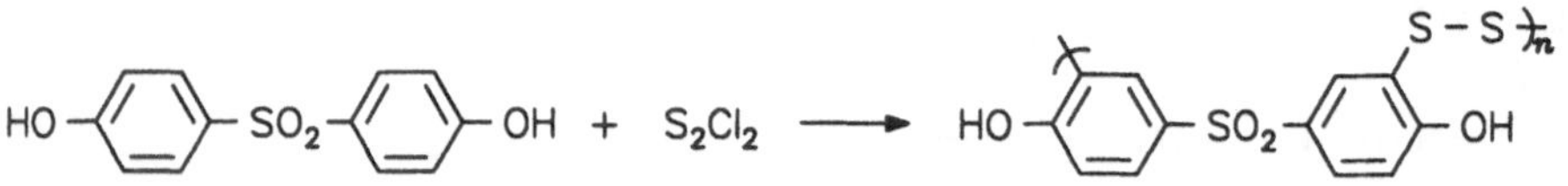

(68)

The poly(amine disulfide) acts as a copper deactivator in a composition containing LDPE and other substances [196]. These copolymers were partly soluble in DMSO and *N*-methyl pyrrolidinone and were stable at 200 °C in air. Copolymers of 3,3′-dinitro-4,4′-dihydroxydiphenylsulfone and of diaminodiphenylsulfone with S_2Cl_2 were also obtained by the same method. Dialkenes and diols can also react with S_2Cl_2 yielding disulfide polymers [20]:

$$n\,H_2C{=}CH-CH{=}CH_2 + S_2Cl_2 \longrightarrow \left[S-CH_2-\underset{Cl}{CH}-\underset{Cl}{CH}-CH_2-S \right]_n \qquad (69)$$

$$HO-R-OH + S_2Cl_2 \longrightarrow \left[O-R-O-S-S \right]_n \qquad (70)$$

Dechlorination with sodium sulfide to remove the labile chlorine, gives products which are useful as additive for oils [197].

2.8 Polysulfides from Disulfenyl Chloride Derivatives

When 4,4′-methylenedisulfenylchloride reacts with dithiols, it forms polydisulfides [198, 199]:

$$Cl-S-R-S-Cl + HS-R'-SH \longrightarrow \left[S-R-S-S-R'-S \right]_n + HCl \qquad (71)$$

where R or R′ can be biphenyl or diphenyl methane. In addition, polymers with R′ being phenyl, hexamethylene and tetramethylene groups were also prepared. When the disulfenylchloride is reacted with lithium metal, poly(biphenyl disulfide) is obtained [198].

2.9 Polysulfides from Cyclic Sulfides by Ionic Mechanism

Optically active and structurally homogenous poly(propylene disulfide) was obtained by asymmetric polymerization of R,S-propylene sulfide:

$$H_3C - CH - CH_2 \longrightarrow \left(\underset{\overset{|}{CH_3}}{CH} - CH_2 - S - S \right)_n + H_3C - CH = CH_2 \quad (72)$$

High specific rotation of disulfide polymer compared to that of poly(propylene sulfide) indicated the presence of chiral S–S linkage [200, 201]. The catalyst used for the polymerization consisted of n-butyllithium and lithium-1-menthoxide or Li alkyl sulfide and lithium-1-menthoxide.

Polymerization of trimethylethylenesulfide, under similar conditions, was accompanied by the evolution of large amounts of trimethylethylene and gave polymers containing tri- and tetrasulfide bonds [202]. Similarly, polymerization of cyclohexenesulfide initiated by lithium alcoholates gave poly(cyclohexenesulfide), poly(cyclohexene disulfide) and poly(cyclohexene trisulfide) [203].

3 Polydiselenides

Due to the presence of metallic selenium, polyselenides are attractive as semiconducting materials. The literature on selenium-containing polymers is scarce. The only polymer that has been extensively studied is poly(methylene selenide). The general routes for preparing polydiselenides include: (i) polycondensation of alkalidiselenides with dihalides, (ii) hydrolysis of alkanediselenocyanates in alkaline medium, and (iii) ring opening polymerization of cyclic diselenides. The first method suffers from the disadvantage of the occurrence of cyclization reactions competing with the polymerization. Only low molecular weight polymers are obtained from the second method. The third method gives high molecular weight polymers in good yields. In spite of these advantages of the third method, because of the difficulty of preparing the cyclic diselenides, the literature on polydiselenides was found to cover mainly the first two polymerization methods. Apart from these three methods, few other methods have been used to make polydiselenides.

3.1 Synthesis of Selenocyanates and Alkali Diselenide (or Ditelluride) Solutions

Although there are many ways of preparing both selenocyanates and alkali diselenide (or ditelluride) solutions, only those methods which are used for making the polymers are discussed here. Alkanediselenocyanates can easily be prepared by condensing the dihalide with potassium selenocyanate in solvents like acetone [204]. Alkali diselenides (or ditellurides) can be prepared in aqueous solution by reacting sodium formaldehyde sulfoxylate with selenium (or tellurium) [205]:

$$4A \quad + \quad HOCH_2SO_2Na \quad + \quad 3NaOH \quad \longrightarrow \quad 2Na_2A_2 \quad (A = Se, Te)$$

$$(73)$$

In liquid NH_3 [206] or in organic solvents like DMF, NMP and HMPA, equal mols of alkali metal and selenium (or tellurium) react to give alkali diselenide (or ditelluride) solution [207, 208]:

$$2Na \quad + \quad mA \quad \longrightarrow \quad Na_2A_m \quad (m = 1, 2) \quad (A = Se, Te) \qquad (74)$$

Lithiumboroethyrate also reacts with selenium (or tellurium) in THF to form lithium diselenide (or ditelluride) [209]:

$$LiBEt_3H \quad + \quad mA \quad \longrightarrow \quad Li_2A_m \quad + \quad 2BEt_3 \quad + \quad H_2$$

$$(m = 1, 2)$$

$$(A = Se, Te) \qquad (75)$$

Since liquid NH_3 is difficult to handle, many reports on the polycondensation have followed the other two procedures.

3.2 Poly(Methylene Diselenide)

Prince et al. [210] found that under dry conditions formaldehyde reacts with sodium selenide to form poly(methylene diselenide):

$$CH_2O \quad + \quad Na_2Se \quad \longrightarrow \quad \{CH_2-Se_2\}_n \qquad (76)$$

Two forms of this polymer are known: 'a' and 'b'. Form 'a' is a crystalline, red-brown solid which can easily be moulded at 70 °C and 5000 psi. Its density and specific resistivity are 3.4 g/cm^3 and 3×10^5 ohm-cm respectively. When heated above 120 °C, form 'a' is transformed into form 'b' which is a black-rubbery polymer having a T_g near room temperature. Form 'b' has a lower specific resistivity (1.6×10^4 ohm-cm) and can be used as a semiconducting material.

Poly(methylene diselenide) can also be prepared by the condensation of dibromomethane and aqueous sodium diselenide [205, 211]. This polymer is found to react with sulfuryl chloride to give equimolar quantities of chloro-methylselenium trichloride and selenium tetrachloride by the cleavage of Se–Se and C–Se bonds [211]:

$$\{CH_2-Se-Se\}_n \; + \; 4n\,SO_2Cl_2 \longrightarrow n\,ClCH_2SeCl_3 + n\,SeCl_4 + \; 4n\,SO_2 \quad (77)$$

3.3 Poly(trimethylene Diselenide)

The 1,3-propanediselenocyanate on hydrolysis with alcoholic KOH yields poly(trimethylene diselenide) [212]:

$$NCSe-(CH_2)_3-SeCN \xrightarrow{\;H^+\;} \{(CH_2)_3-Se_2\}_n \qquad (78)$$

$$\uparrow {\scriptstyle -nH_2C=CH-CH_3}$$

$$\{(CH_2)_3-Se\}_n$$

which when heated in a solution above 60 °C degrades to give 1,2-diselenolane monomer [213]:

$$\{(CH_2)_3-Se_2\}_n \; \xrightleftharpoons{\;60\,°C\;} \; \overset{Se}{\underset{Se}{\bigsqcup}} \qquad (79)$$

After standing for several days at room temperature, the poly(trimethylene diselenide) slowly precipitated out from the solution. Poly(trimethylene selenide) on degradation also forms the diselenide polymer with the evolution of propylene (Eq. 78) [212].

3.4 Other Aliphatic Diselenide Polymers

Like poly(trimethylene diselenide), other higher hydrocarbon diselenide polymers were prepared either by hydrolysis of respective selenocynate

$$NCSe-(CH_2)_x-SeCN \xrightarrow{\;H^+\;} \{(CH_2)_x-Se_2\}_n \qquad (80)$$

(x = 4 [214, 215], 5 [216] and 6 [217]) or by the condensation of respective dihalide with alkali diselenide (x = 2–20) [205].

Poly(tetramethylene diselenide) in chloroform solution at 60 °C or in the presence of diffused sunlight degrades to form 1,2-diselenane [218] and when heated strongly it forms selenolane [214]:

$$\{(CH_2)_4-Se_2\}_n \; \xrightarrow[\substack{or\,h\nu \\ \Delta}]{60\,°C} \; \overset{Se}{\underset{Se}{\bigcirc}} \qquad (81)$$

$$\overset{}{\underset{Se}{\bigcirc}} \; + \; Se \qquad (82)$$

Poly(pentamethylene diselenide) on heating forms selenane [216]:

$$\left(\!\!\left(CH_2\right)_5\!-\!Se_2\right)_{\!\overline{n}} \quad \xrightarrow{\ \Delta\ } \quad \text{(ring)}{}^{Se} \quad + \quad Se \tag{83}$$

Similarly, when poly(hexamethylene diselenide) was heated above 250 °C it degrades to form 2-methylselenane and selenium [217]:

$$\left(\!\!\left(CH_2\right)_6\!-\!Se_2\right)_{\!\overline{n}} \quad \xrightarrow{\ \Delta\ } \quad \text{(ring)}_{Se\ \ CH_3} \quad + \quad Se \tag{84}$$

Reactions of poly(tetramethylene diselenide) and poly(pentamethylene diselenide) with Br_2 yielded 1,1-dibromoselenolane [214] and 1,1-dibromoselenane [216] respectively. Fredga [219, 220] prepared diselenide polymers having functional groups such as acid groups by the ring opening polymerization of 1,2-diselenane-3,6-dicarboxylic acid:

$$\text{(ring structure)} \quad \xrightarrow{\ \Delta\ } \quad \left(\!Se-\underset{\underset{COOH}{|}}{CH}-CH_2-CH_2-\underset{\underset{COOH}{|}}{CH}-Se\right)_{\!\overline{n}} \tag{85}$$

By condensing dibromopimelic acid and sodium diselenide, Fredga and Styrman [221] reported another acid group containing polymer. The acid hydrolysis of di(selenocyanato)derivatives of both the *rac*- and *meso*-dibromopimelic acids also yielded in each case the polymeric diselenides. Polymers containing both selenium and sulfur in the backbone have been reported by Bergson et al. [222]:

$$\underset{Se}{\overset{S}{\diagdown}}\!\!\!\diagup\!\!-COOH \quad \xrightarrow{heat > m.p.} \quad \begin{cases} \left(CH_2-\underset{\underset{COOH}{|}}{CH}-CH_2-Se-S\right)_{\!\overline{n}} \\[2em] \left(CH_2-\underset{\underset{COOH}{|}}{CH}-CH_2-Se-Se-CH_2-\underset{\underset{COOH}{|}}{CH}-CH_2-S-S\right)_{\!\overline{n}} \end{cases}$$

$$\tag{86}$$

On heating the 1-thia-2-selenolane-4-carboxylic acid above its melting point, it polymerizes, for which two alternative structures have been proposed.

Similarly, 1-thia-2-selenolane-4-aminomethylhydrochloride (7)

$$\underset{S}{\overset{Se}{\diagdown}}\!\!\!\diagup\!\!-CH_2-NH_3^+\,Cl^-$$

7

also polymerizes on heating at 120 °C [223].

3.5 Aromatic Diselenide Polymers

Phenylene diselenide polymers were first prepared via the bis-Grignard reagents and through the bis-diazonium salts [224]. While poly(m-phenylene diselenide) (mol. wt. = 1450) is amorphous [224], the p-phenylene diselenide polymer exists as a crystalline tetramer [224, 225]. Later, o-phenylene diselenide polymer was reported by Sandman et al. by the polycondensation of o-dichlorobenzene with sodium diselenide [226]. The end groups were found to be the unreacted chloride groups. Poly(o-phenylene diselenide) on reduction with sodium-borohydride followed by treatment with thiophosgene forms 4,5-benzo-1,3-dis-elenole-2-thione. When metal chlorides like nickel chloride was used instead of thiophosgene, it forms tetra-n-butylammonium nickel bis(benzene-1,2-dis-elenolates) [226]:

$$\text{(87)}$$

$$[(C_4H_9)_4N][Ni(C_6H_4Se_2)_2] \qquad \text{(88)}$$

Similarly, poly(9,10-diseleno anthracene) was made by condensing 9,10-dib-romoanthracene with sodium diselenide [206]. Okamoto et al. [225] prepared poly(4,4'-biphenylene diselenide) in two forms having m.p. 145 °C and 190 °C, mass spectrometry indicating the component melting at 145 °C to be the cyclic dimer. Poly(m-xylene diselenide) was prepared by the hydrolysis of the corres-ponding selenocyanate. The cyclic diselenide which was also formed during the reaction forms the same polymer on heating [204]:

$$\text{(89)}$$

It appears remarkable that benzylic and poly(p-phenylene diselenide poly-mers are able to incorporate additional selenium atoms in to the chain through a simple melt "alloying" process, thus giving products whose structures are intermediate between organic and inorganic polymers [224, 227]. Recently, poly(1,1'-binaphthyl-2,2'-diselenide) was prepared along with other compounds when 2,2'-dilithio-1,1'binaphthyl was reacted with selenium followed by aerial oxidation [228].

4 Polyditellurides

Like polyselenides, the main theme of making polytellurides is also electrical conductivity. Hence, the main focus has been in making polymers with conjugated backbone for better conductivity.

4.1 Poly(Methylene Ditelluride)

Morgan and Drew [229] reduced bistrichlorotelluro(IV)methane with potassium metabisulfite and got red-brown ditelluride with a m.p. of 50–90 °C which was proposed by Berry et al. [230] to be a dimer, $(CH_2Te_2)_2$. When this reaction was repeated by Dirk et al. [231], they confirmed the formation of a red powder which subsequently turned into a grey solid. The X-ray diffraction results showed both red and grey solids to be amorphous contrary to the crystalline red solid reported previously [229, 230]. With X-ray evidence, the red or grey powders have been proposed to be poly(methylene ditelluride):

$$
\begin{array}{c}
\text{H}\diagdown \quad \diagup \text{TeCl}_3 \\
\text{C} \\
\text{H}\diagup \quad \diagdown \text{TeCl}_3
\end{array}
\xrightarrow{\text{K}_2\text{S}_2\text{O}_5}
\quad \text{+}\!\!\left(\,CH_2-Te_2\,\right)\!\!\overline{}_n
\tag{90}
$$

This polymer showed one endothermic transition at 55 °C and several exothermic transitions at 100, 130, 175, 190 and 230 °C. The latter was the largest and corresponded to a decomposition reaction as determined by TGA. At 55 °C the polymer becomes tacky and can be drawn into fibres or rolled into sheets. Fiber drawn samples also did not show any X-ray pattern. On degradation the polymer formed cyclic compounds which were confirmed by mass spectroscopy and IR. While the IR of the degraded products was almost superimposable on the polymer, the mass spectrum showed peaks corresponding to $(TeCH_2Te)$ (m/z 272) and $(TeCH_2TeTeCH_2)$ (m/z 412). The formation of cyclic products may also be ascribed to the evaporation of the insoluble compounds mixed with the polymer. Poly(methylene ditelluride) in $CHCl_3$ reacts with Cl_2 to form tellurium tetrachloride and chlorine derivatives of the organotellurium compounds:

$$
\text{+}\!\!\left(\,CH_2-Te_2\,\right)\!\!\overline{}_n \;+\; Cl_2 \;\longrightarrow\; CH_2(TeCl_3)_2 \;+\; TeCl_4 \;+\; ClCH_2TeCl_3 \tag{91}
$$

The polymer was found to be an insulator ($R > 10^{10}$ ohm-cm). When the sample was heated above 60 °C, evolution of small fragments occurs with a concomitant dramatic drop in resistivity, cross-linking and generation of unpaired spins. The ESR of heat treated samples showed a broad 800 G wide line whereas the untreated sample showed no resonance signals. Another study [232] on poly(methylene ditelluride), prepared by using the boroethyrate procedure, revealed that its conductivity increased by four to nine orders of magnitude on doping with bromine or iodine. Also, the polymer was found to be very much crystalline contrary to the previous report [231]. Poly(methylene ditelluride) on reduction followed by treatment with either methyl sulfate and bromine [232] or with methylene bromide or iodide [233] yielded poly(methylene telluride):

$$\pmb{\{}CH_2-Te_2\pmb{\}_n} \xrightarrow{NaBH_4} x\,Na^+Te^-CH_2Te^-Na^+ \xrightarrow[Br_2]{Me_2SO_4} \pmb{\{}CH_2-Te\pmb{\}_n} \quad (92)$$

4.2 Other Aliphatic Ditelluride Polymers

Nogami et al. [205] prepared many aliphatic ditelluride polymers by condensing dihalides (x = 2–20) with alkali metal ditelluride:

$$Br-(CH_2)_x-Br \;+\; Na_2Te_2 \longrightarrow \pmb{\{}(CH_2)_x-Te_2\pmb{\}_n} \quad (93)$$

From fluorescence X-ray measurements, the percentage of bromine present in the polymers was found to be less than 1% of that of tellurium, suggesting that the end groups are bromides [209]. Like any other polycondensation reactions, formation of cyclics also occurs along with that of the polymer. For example, the reaction of n-butyldiiodide with sodium ditelluride results in the formation of the polymer along with the six-membered cyclic ditelluride [234]:

$$I-(CH_2)_4-I \;+\; Na_2Te_2 \longrightarrow \underset{Te-Te}{\bigcirc} \;+\; \pmb{\{}(CH_2)_4-Te_2\pmb{\}_n} \quad (94)$$

4.3 Aromatic Ditelluride Polymers

Poly(p-phenylene ditelluride) and poly(biphenylene ditelluride) were synthesized by Nogami et al. [235]:

$$(95)$$

Similarly, the biphenylene derivative was prepared by starting with 4,4'-dibromodiphenyl. The X-ray diffraction of these two polymers proved them to be amorphous [235]. Conversely, other work [236] reported the melting point of poly(p-phenylene ditelluride) to be 158 °C with decomposition around 390 °C. Both phenyl and biphenylene ditelluride polymers exhibited photoconductivity upon irradiation by near-infrared light at around 1000 nm. The electrical conductivities of these polymers increased from insulators (10^{-12} S cm^{-1}) to semiconductors $(10^{-4}-10^{-6}\,\mathrm{S\,cm^{-1}})$ upon doping with Br_2 or I_2 [235]. Poly(p-phenylene ditelluride) was also prepared by condensing p-dibromobenzene with sodium ditelluride [207]. Syntheses of other aromatic polymers such as poly(1,1'-biphenyl-2,2'-ditelluride) [237], poly(1,1'-binaphthyl-2,2'-ditelluride) [228] and poly(p-xylene ditelluride) [232] have also been achieved [193]. Sadekov et al. also reported the synthesis of benzoditelluorole [238] and telluroanthrene [239] from poly(o-phenylene ditelluride).

$$(96)$$

$$(97)$$

5 Conclusions

One of the unique characteristics of polysulfides is their unusually high molecular weights which can be obtained through condensation polymerization. Due to the high molecular weight they are insoluble in common organic solvents and hence the determination of molecular weight and its distribution is not possible. Studies on molecular weight distribution of liquid polysulfide polymers have produced contradicting results. The insolubility of high molecular weight polymers also prohibits the use of spectral characterization. Modern techniques like mass spectrometry may provide detailed information on the distribution of polysulfide linkages in these polymers. Due to disulfide–disulfide and disulfide–thiol interchange reactions, polysulfide polymers provide an interesting case for relaxation studies.

Like polysulfide polymers, most of the polydiselenides and polyditellurides, being insoluble, have not been properly characterized, and their molecular weights are not available. Controversy exists on the crystalline nature of polyditellurides and formation of cyclics during the polycondensation to form polyditellurides. Due to the presence of metals, these polymers are attracting attention for their conducting properties. However detailed investigations are needed to establish structure-conductivity relationships for exploiting their potential.

Due to the weak-bond energies of group VIA linkages, knowledge of the thermal and photo stabilities of these polymers, particularly for diselenides and ditellurides, are needed for their successful applications. None of these polymers have been studied adequately to establish their degradation behaviour, including the reactivities of the generated radicals and the associated energetics. A comparative study of group VIA polymers may provide interesting avenues for learning more about the chemistry of alkoxy, thiyl, selenyl and telluryl radicals. The peroxide polymers show unusual exothermicity, and such energetic study should be carried out on other structurally similar polymers in the group to obtain a comprehensive view on their structure-property relationships.

To summarize, it may be said that the disulfide, diselenide and ditelluride polymers are not only attractive industrially important materials but they also exhibit interesting chemistry judging from what little is known about them.

6 References

1. Kishore K, Ravindran K (1982) Macromolecules 15: 1638
2. Mukundan T, Kishore K (1987) Macromolecules 20: 2382
3. Mukundan T, Kishore K (1989) Macromolecules 22: 4430
4. Kishore K, Mukundan T (1986) Nature 324: 130
5. Mukundan T, Bhanu VA, Kishore K (1989) J Chem Soc, Chem Commun 12: 780
6. Jayanthi S, Kishore K (1993) Macromolecules (in press)
7. Mukundan T, Kishore K (1989) J Polym Sci, Polym Lett Ed 27: 455
8. Mukundan T, Annakutty KS, Kishore K (1992) Fuel (in press)
9. Mukundan T, Kishore K (1990) Prog Polym Sci 15: 475
10. Kishore K, Gayathri V, Ravindran K (1981) J Macromol Sci Chem A16: 1359
11. Mukundan T, Kishore K (1991) Current Sci. 60: 355
12. Bhanu VA, Kishore K (1991) Chem. Rev. 91: 99
13. Fettes EM (1961) in: Kharasch N (ed) Organic Sulfur compounds, Vol. I. Pergamon press, Oxford, p 266
14. Goethals EJ (1968) J Macromol Sci -Revs Macromol Chem 73: 1968
15. Berenbaum MB (1968) in: Encylopedia of Chemical Technology, Vol 16. Interscience, New York, p 253
16. Gobran RH, Berenbaum MB (1969) in: Kennedy JP, Tornquist EGM (eds) High Polymers, Vol. XXIII. Interscience publishers, p 805
17. Ellerstein SM, Bertozzi ER (1982) in: Encyclopedia of Chemical Technology, Vol. 18. Wiley-Interscience, New York, p 814
18. Gaylord NG (ed) (1962) Polyethers, Part III: Polyalkylene sulfides and other polythioethers. Interscience, New York, p 1
19. Tobolsky AV (ed) (1968) The Chemistry of Sulphides. Wiley-Interscience publishers, New York, p 221
20. Vietti DE (1989) in: Allen G, Bevington JC (eds) Comprehensive Polymer Science, Vol 5. Pergamon press, p 533
21. Spassky N, Sepulchre M, Sigwalt P (1992) in: Kricheldorf HR (ed) Handbook of Polymer Synthesis, Part B. Marcel Dekker, New York, p 991
22. Mortillaro L, Russo M (1973) in: Klayman DL, Gunther WHH (eds) Organic Selenium Compounds: Their Chemistry and Biology. Wiley-Interscience, New York, p 815
23. Patrick JC, Mnookin NM (1927) Br. Patent No. 302270
24. Fettes EM, Jorczak JS (1950) Ind Eng Chem 42: 2217
25. Blansma JJ (1901) Rec Trav Chim 20: 121
26. Lenz RW (1971) Organic Chemistry of Synthetic High Polymers. Interscience, New York, p 154
27. Joshi UR, Limaye PA (1984) Biovigyanam 10: 185
28. Trochimczuk W (1970) Polimery 15: 349
29. Tokarzewski L, Plonka S (1969) Polimery 14: 596

30. Annenkova VZ, Khaliullin AK, Bugun LG, Korobogatova VI, Il'icheva LN, Voronkov MG (1985) Zh Prikl Khim (Leningrad) 58: 1187
31. Spielberger G (1963) in: Methoden der Organischen Chemie, Band XIV/2. Georg Thieme Verlag, Stuttgart, p 591
32. Bertozzi ER (1957) US Patent No. 2796325
33. Voronkov MG, Annenkova VK, Khaliullin Ak, Antonik LM (1979) Vysokomol Soed 21B: 235
34. Martin Jr SM, Patrick JC (1936) Ind Eng Chem 28: 1144
35. Sander SR, Karo W (1930) in: Polymer Synthesis III. Academic press, New York, p 68
36. Patrick JC (1934) US Patent No. 1950744
37. Bertozzi ER, Davis FO, Fettes EM (1956) J. Polym. Sci. 19: 17
38. Averko-Antonovich LA, Kirpichnikov PA, Nigmatullina FG (1986) Izv Vyssh Uchebn Zved, Khim Khim Teknol 29: 104
39. Kshirsagar SN (1982) Pop Plast 27: 3
40. Todorova D, Mladenov Iv, Markov M, Todorov St (1984) J Mol Struct 114: 421
41. Khristova D, Markov M, Mladenov I (1980) Plast Kautsch 27: 371
42. Todorova D, Markov M, Mladenov I (1980) God Vissh Khim-Tekhnol Inst, Burgas, Bulg 16: 28
43. Markov M, Todorov T, Mladenov I (1986) Polimery 31: 165
44. Shutov AA, Vaiman EYa (1973) Izv Vyssh Ucheb Zaved Khim Khim Tekhnol 16: 1736
45. Fettes EM, Jorczak JS, Panek JR (1954) Ind Eng Chem 46: 1539
46. Fettes EM (1954) US Patent No. 2676165
47. Fettes EM (1952) US Patent No. 2606173
48. Patrick JC (1936) Trans Faraday Soc 32: 347
49. Trilliat JJ, Tertian R (1944) Compt Rend 219: 395
50. Tertian R (1946) Rev Gen Caoutchonc 23: 245
51. Dawson IM, Mathieson AM, Robertson JM (1948) J Chem Soc 322
52. Dawson IM, Robertson JM (1948) J Chem Soc 1256
53. Donohue J, Schomaker V (1948) J Chem Phys 16: 92
54. Foss O (1950) Acta Chem Scand 4: 404
55. Koch HP (1949) J Chem Soc 394
56. Baer JE, Carmack M (1949) J Am Chem Soc 71: 1215
57. Schotte L (1956) Arkiv fur kemi 9: 361
58. Trofimov BA, Nedolya NA, Komel'kova VI, Khil'ko MYa (1986) Zh. Pr. Khim. 59: 2382
59. Wulff G, Schulze I (1978) Angew Chem Int Ed Engl 17: 537
60. Meyer VE, Dergazarian TE (1984) US Patent No. 4,438,259
61. Dergazarian TE (1986) US Patent No. 4,607,078
62. Meyer VE, Dergazarian TE (1986) US Patent No. 4,608,433
63. Hefner Jr RE (1987) US Patent No. 4692700
64. Te Grotenhuis TA, Swart GH (1953) US Patent No. 2631994
65. Ryden LL (1946) US Patent NO. 2406260
66. Novoselok FB, Sokolov VN, Apukhtina NP, Shlyakhter RA (1965) Vysokomol Soed 7: 1726
67. Tenc-Popovic IE, Rekalic VJ, Radosavlhevic SD (1987) J Serb Chem Soc 52: 89
68. Radosavljevic SD, Stasic L, Rekalic VJ, Tenc-Popovic ME (1979) Glas Hem Drus Beograd 44: 631
69. Radosavljevic SD, Rabrenovic MD, Rekalic VJ, Tenc-Popovic ME (1978) Glas Hem Drus Beograd 43: 105
70. Radosavljevic SD, Tenc-Popovic ME, Rekaliv VJ (1970) Glas Hem Drus Beograd 35: 397
71. Rekalic VJ, Radosavljevic SD, Tenc-Popovic ME (1970) J Polym Sci Part A-1. 8: 3259
72. Rekalic VJ, Tenc-Popovic ME, Radosavljevic SD, Zozuk J (1982) Glas Hem Drus Beograd 47: 637
73. Rekalic VJ, Tenc-Popovic ME, Radosavljevic SD (1980) J Polym Sci Polym Chem Ed 18: 2033
74. Radosavljevic SD, Tenc-Popovic ME, Stasic L, Rekalic VJ (1976) Glas Hem Drus Beograd 41: 115
75. Patrick JC, Ferguson HR (1945) US Patent No. 2466963
76. Davis FO (1952/53) US Patent No. 2715635
77. Patrick JC, Ferguson HR (1944) US Patent No. 2402977
78. Jorczak JS, Fettes EM (1951) Ind Eng Chem 43: 324
79. Kirchhof F (1957) Kautschuk u Gummu 10: 176, 227, 232
80. Rafikov SR, Ionov VI, Aleyev RS, Pestryayev YeM, Pancheshnikova RB, Dobrikov AL, Danilov, VT (1978) Vysokomol Soyed 20A: 516

81. Rama Rao M, Radhakrishnan TS (1985) J Appl Polym Sci 30: 855
82. Colodny PC, Tobolsky AV (1959) J Appl Polym Sci 2: 39
83. Tobolsky AV, Macknight WJ, Takahashi M (1964) J Phys Chem 68: 787
84. Genkin AN, Nasonova TP, Poddubnyi IYa, Shlyakhter RA (1962) Vysokomol Soed 4: 1088
85. Shlyakhter RA, Apukhtina NP, Nasonova TP (1963) Dokl Akad Nauk SSSR 149: 345
86. Fettes EM, Mark H (1963) J Appl Polym Sci 7: 2239
87. Shlyakhter RA, Ehrenburg EG, Nasonova TP, Piskareva EP (1965) in: IUPAC International
    Symposium on Macromolecular Chemistry, Prague.
88. Hayashi M, Shiro Y, Murata H (1966) Bull Chem Soc Jpn 39: 1857
89. Ohsaku M, Imamura A (1984) Polym Commun 25: 251
90. Hayashi M, Shiro Y, Murata H (1966) Bull Chem Soc Jpn 39: 1861
91. Isaacs LG, Fox RB (1965) J Appl Polym Sci 9: 3489
92. Bertozzi ER (1950) US Patent No. 2527378
93. Patrick JC (1949) US Patent No. 2469404
94. Okita T (1945) US Patent No. 2378576
95. Okita T (1948) J Soc Rubber Ind Japan 21: 35
96. Okita T (1948) J Soc Rubber Ind Japan 21: 77
97. Ballard SA, Morris RC, Vanwinkle JL (1949) US Patent No. 2484369
98. Bennett GM (1922) J Chem Soc 2139
99. Bennett GM, Hock AL (1927) J Chem Soc 477
100. Morris RC, Vanwinkle JL (1950) US Patent No. 2518245
101. Richter F, Augustine GB, Koft E, Reid EE (1952) J Am Chem Soc 74: 4076
102. Richter FP, Reid EE (1952) US Patent No. 2582605
103. Okita T (1943) US Patent No. 2332869
104. Sumitomo Electric Wire & Cable Works (1938) Brit Patent No. 491363
105. Sumitomo Electric Wire & Cable Works, (1938) Brit Patent No. 491309
106. Abrams JT, Andrews KJM, Woodward FN (1959) J Polym Sci 41: 255
107. Andrews KJM, Rosser RJ, Woodward FN (1959) J Polym Sci 41: 231
108. Farbenind IG (1953) Ger Patent No. 612665
109. Twiss DF, Neale AET (1936) US Patent No. 2056026
110. Patrick JC (1944) US Patent No. 2363614
111. Murayama K, Morimoto S, (1967) Chem Ind (London) 10: 402
112. Murayama K, Kato Y, Morimoto S (1967) Bull Chem Soc Jpn 40: 2645
113. Schimmelschmidt K (1962) Angew Chem 74: 975
114. Kwan TC (1962) US Patent No. 3054781
115. Davis FO, Fettes EM (1948) J Am Chem Soc 70: 2611
116. Tobolsky AV, Leonard F, Roesser GP (1948) J Polym Sci 3: 604
117. Davis FO (1953) US Patent No. 2657198
118. Davis FO (1955) US Patent No. 2715635
119. Affleck JG, Dougherty G (1950) J Org Chem 15: 865
120. Barltrop JA, Heyes PM, Calvin M (1954) J Am Chem Soc 76: 4348
121. Reed LJ, Niu CI (1955) J Am Chem Soc 77: 416
122. Schotte L (1956) Arkiv fur Kemi 9: 441
123. Thomas RC, Reed LR (1956) J Am Chem Soc 78: 6148
124. Schotte L (1956) Arkiv Kem 9: 309
125. Schoberl A, Grafje H (1958) Liebigs Ann Chem 614: 66
126. Cragg RH, Weston AF (1973) Tetrahedron Lett. 655
127. Gronowitz S, Lidert Z (1980) Chem Scripta 16: 97
128. Lidert Z, Gronowitz S (1980) Chem Scripta 16: 102
129. Field L, Stevens WD, Lippert ELJr (1961) J Org Chem 26: 4782
130. Dainton FS, Davies JA, Manning PP, Zahir SA (1957) Trans Faraday Soc 53: 813
131. Houk J, Whitesides GM (1989) Tetrahedron 45: 99
132. Tobolsky AV, Leonard F, Roeser GP (1948) J Polym Sci 3: 604
133. Davis FO, Fettes FM (1948) J Am Chem Soc 70: 2611
134. Dainton FS, Ivin KJ, Walmsley DAG (1960) Trans Faraday Soc 56: 1784
135. Tobolsky AV, Meltzer TH (1955) US Patent No. 2728750
136. Tobolsky AV, Baysal B (1953) J Am Chem Soc 75: 1757
137. Stockmayer WH, Howard RO, Clarke JT (1953) J Am Chem Soc 75: 1756
138. Hallensleben ML (1977) J Polym Sci Polym Lett 15: 619
139. Hallensleben ML (1977) Eur Polym J 13: 437

140. Krespan CG (1968) in: Tobolsky AV (ed) The chemistry of sulfides. Interscience, New York, p 211
141. Fedotova OYa, Grozdov AG, Koesnikov GS (1967) Vysokomol Soed 9B: 456
142. Ghafoor A, Senior JM, Still RH, West GH (1974) Polymer 15: 577
143. Cho H, Kim Sc, Kwon YH, Chung WK (1980) Hanguk Sumyu Konghakhoe Chi 17: 90
144. Hiroaki T, Masotoshi I (1965) Kogyo Kagaku Zasshi 68: 1752
145. Bruma M, Dumitrescu GB (1980) Rom Patent No. 70228
146. Dumitrescu G, Dumitrescu E, Chiriac C (1980) Rom Patent No. RO 71423
147. Baxter RL, Glover SSB, Gordon EM, Gould RO, Mckie MC, Ian Scott A, Walkinshaw MD (1988) J Chem Soc, Perkin Trans I 365
148. Harris FW, Eury RP Polym Prepr 449
149. Cho H, Him SC, Kwon YH, Lee SY (1980) Yongu Pogo-Yongnam Taehakkyo Kongop Kisul Yongusu 8: 49
150. Fujimoto A, Endo T, Okawara M (1974) Makromol Chem 175: 3597
151. Bochkareva LN, Kattsyna II, Bochkarev VV, Bondaletov VG (1987) USSR Patent No. SU 1291592
152. Manecke G, Willie WE (1970) Makromol Chem 133: 61
153. Patrick JC (1944) US Patent No. 2363614
154. Patrick JC (1939) US Patent No. 2142145
155. Thiokol Corp. (1937) Brit Patent No. 465786
156. Thiokol Corp. (1937) Brit Patent No. 464356
157. Nummy WR (1958) US Patent NO. 2866776
158. Marvel CS, Olson LE (1957) J Am Chem Soc 79: 3089
159. Hay AS (1966) US Patent No. 3294760
160. Hay AS (1963) Fr. Patent No. 1337285
161. Kaneko M (1976) Japan Kokkai 7631796
162. Montaudo G, Bruno G, Maravigna P, Bottino F (1974) J Polym Sci, Polym Chem 12: 2881
163. Weissflog E (1983) Phosphorous Sulfur 15: 27
164. Weissflog E, Schmidt M (1979) Phosphorous Sulfur 6: 453
165. Weissflog E (1983) Phosphorous Sulfur 14: 233
166. Bernhart DN (1972) US Patent No. 3686329
167. Karabinos JV, Yiannios CN (1970) US Patent No. 3513088
168. Goethals EJ, Sillis C (1968) Makromol Chem 119: 249
169. Rassch MS (1979) J Org Chem 44: 2629
170. Casa CD, Bizarri PC, Nuzziello S (1985) J Polym Sci, Polym Lett Ed 23: 323
171. Bizarri PC, Casa CD, Fiorni M (1988) J Polym Sci Polym Chem 26: 255
172. Bonsignore PV, Marvel CS, Bannerjee S (1960) J Org Chem 25: 237
173. Kobayashi N, Osawa A, Fujisawa T (1973) J Polym Sci Polym Lett 11: 225
174. Kobayashi N, Osawa A, Fujisawa T (1975) J Polym Sci Polym Chem 13: 2863
175. Kobayashi N, Fujisawa T (1972) J Polym Sci Part A-1 10: 2591
176. Kobayashi N, Fujisawa T (1973) J Polym Sci Polym Chem 11: 545
177. Fitch RM, Helgeson DC (1969) J Polym Sci Part C 22: 1101
178. Fujisawa T, Kakutoni M (1970) J Polym Sci, Part B 8: 19
179. Rosenthal NA (1962) US Patent No. 3070569
180. US Patent No. 3843600 (1970)
181. US Patent No. 3845013 (1970)
182. Robinson RE, Mueller MW (1964) US Patent No. 3119795
183. Shim KS (1973) US Patent No. 3737461
184. Shim KS (1973) US Patent No. 3742058
185. Paushkin YaM, Losev YuP, Karakozova ET, Isakovich VN (1973) Vysokomol Soed 15A: 2496
186. Losev YuP, Paushkin YaM, Isakovich VN (1974) Vysokomol Soed 16A: 2502
187. Paushkin YaM, Loseva LP, Dmitrieva LP, Losev YuP, Isakovich VI (1973) Dokl Akad Nauk SSSR 210: 865
188. Losev YuP, Loseva LP, Paushkin YaM. Isakovich VN (1980) USSR Patent No. 734183
189. Loseva LP, Metelitsa DI, Paushkin YaM, Losev YuP (1974) Dokl Akad Nauk SSSR 219: 653
190. Loseva LP, Paushkin YaM, Losev YuP, Isakovich VN (1973) Dokl Akad Nauk SSSR 213: 129
191. Ksenofonto MA, Volod'ko LV, Elistratov GN, Paushkin YaM, Losev YuP, Ugolev II, Isakovich VN (1975) Dokl Akad Nauk BSSR 19: 714
192. Losev YuP, Isakovich VN, Paushkin YaM, Loseva LP, Ksenofontov MA, Volod'ko LV (1980) Vysokomol Soed 22A: 607

193. Paushkin YaM, Lumin AF, Losev YuP (1989) Dokl Akad Nauk SSSR 307: 1415
194. Shrinskaya LP, Komarov VS, Loseva LP, Bel'skaya RI (1981) Vestsi Akad Navuk BSSR, Ser Khim Navuk 5: 121
195. Nesterovich VN, Petryaev EP, Shadyro OI (1984) Vestsi Akad Navuk BSSR, Ser Khim Navuk 2: 88
196. Losev YuP, Nesterovich VN, Petryaev EP, Firsov YuI, Shadyro OI, Yakubovskaya EP (1984) USSR Patent No. SU 1079654
197. Bolle J, Dabir A (1981) US Patent No. 4284520
198. Cameron GG, Hogg DR, Stachowiak SA (1975) Makromol Chem 176: 9
199. Cameron GG, Stachowiak SA (1975) Makromol Chem 176: 1523
200. Aliev AD, Solomantina IP, Krentsel BA (1976) Tezisy Dokl-Vses Knof "Stereokhim Konform Anal Org Neftekhim Sint", 3rd p 27
201. Aliev AD, Alieva SL, Krentsel BA (1980) Vysokomol Soed 22A: 1171
202. Aliev AD, Solomantina IP, Zhumabaev Zh, Krenstel BA (1976) in: Bakhtalze IU (ed) Tezisy Dokl Nauchn Sess Khim Tekhnol Org Soedin Sery Sernistykh Neftei p 224
203. Aliev Ad, koshevnik AYu, Gasanov FD (1979) Vysokomol Soed 21A: 1340
204. Mitchell RG (1976) Can J Chem 54: 238
205. Mikawa R, Nogami T, Hasegawa T (1986) Jpn Kokai Tokkyo Koho JP 6128527
206. Xerox Corp. (1974) Neth Appl 7314628
207. Sandman DJ, Stark JC, Acampora LA, Gagne (1983) Organometallics 2: 549
208. Sandman DJ, Stark JC, Foxman BM (1982) Organometallics 1: 739
209. Nogami T, Hasegawa T, Inoue K, Mikawa H (1984) Polym Commun 25: 329
210. Prince M, Bremer B (1967) J Polym Sci, Part B 5: 847
211. Paetzold R, Kanust D (1970) Z Chem 10: 269
212. Morgan GT, Burstall FH (1930) J Chem Soc 1497
213. Bergson G (1962) Ark Kemi 19: 195
214. Morgan GT, Burstall FH (1929) J Chem Soc 1096
215. Bergson G (1958) Ark Kemi 13: 11
216. Morgan GT, Burstall FH (1929) J Chem Soc 2197
217. Morgan GT, Burstall FH (1931) J Chem Soc 173
218. Brown JR, Gillman GP, George MH (1967) J Polym Sci, Part A-1 5: 903
219. Fredga A (1933) Ark Kemi Mineral Geol 11B: 1
220. Fredga A (1935) Uppsala Univ Arsskrift 5: 125
221. Fredga A, Styrman K (1959) Ark Kemi 14: 461
222. Bergson G, Biezais A (1961) Ark Kemi 18: 143
223. Bergson G (1962) Ark Kemi 19: 75
224. Gunther WHH, Salzman MB (1972) Ann N Y Acad Sci 192: 25
225. Okamato Y, Yano T, Homsany R (1972) Ann N Y Acad Sci 192: 60
226. Sandman DJ, Allen GW, Acampora LA, Stark JC, Jansen S, Jones MT, Ashwell GJ, Foxman BM (1987) Inorg Chem 26: 1664
227. Gunther WHH (1972) US Patent No. 3671467
228. Murata S, Suzuki T, Yanagisawa A, Suga S (1991) J. Heterocyclic Chem 28: 433
229. Morgan GT, Drew HDK (1925) J Chem Soc 531
230. Berry FJ, Smith BC, Jones CHW (1976) J Org Metal Chem 110: 201
231. Dirk CW, Nalewajek D, Blanchet GB, Schaffer H, Moraes F, Boysel RM, Wudl F (1985) J Am Chem Soc 107: 675
232. Nogami T, Tasaka Y, Inoue K, Mikawa H (1985) J Chem Soc Chem Commun 5: 269
233. Sadekov ID, Rivkin BB, Maksimenko AA, Minkin VI (1986) Zh Org Khim 22: 2615
234. Singh HB, Khanna PK (1988) J Org Metal Chem 338: 9
235. Hasegawa T, Nogami T, Mikawa H (1984) J Polym Sci, Polym Lett 22: 365
236. Mitsubishi Chemical Industries Co. Ltd (1985) Japan Patent No. 6013814
237. Sadekov ID, Rivkin BB, Maslakov AG, Minkin VI (1987) Khim Geterotsikl Soedin 3: 420
238. Rivkin BB, Sadekov ID, Minkin VI (1988) Khim Geterotsikl Soedin 8: 1144
239. Sadekov ID, Rivkin BB, Gadznieva PI, Minkin VI (1990) Khim Geterotsikl Soedin 1: 137

Editor: Prof. T. Saegusa
Received June 1, 1993

Thermal Discoloration Chemistry
of Styrene-*co*-Acrylonitrile

D.B. Priddy
Dow Plastics, Midland, MI 48667, USA

Styrene-*co*-acrylonitrile resins (SAN) are important commercial thermoplastic materials. However, SAN discolors during molding, the extent of which is dependent upon the AN content of the polymer. This paper reviews the literature dealing with the microstructure of the SAN backbone, the oligomers formed during its manufacture, and their possible roles in SAN thermal discoloration. The recent application of GPC-UV/vis analysis to the SAN discoloration problem has shown that both monomer sequence distribution in the backbone and small molecules are involved in the formation of chromophores and indicates that the main backbone chromophore resides on the chain end. The product of reactivity ratios (calculated from monomer sequence distribution data) of commercially produced SAN copolymers is much higher than ever reported in the literature. Possible mechanisms for this observation and of SAN backbone discoloration are discussed.

1 Introduction

The copolymerization of styrene (S) and acrylonitrile (AN) has been studied in much detail over the last 50 years. Styrene-*co*-acrylonitrile (SAN) resins have improved toughness as well as heat and solvent resistance properties over polystyrene and have therefore become very important commercial thermoplastics. The copolymerization of S and AN is carried out commercially utilizing three types of free radical polymerization processes; i.e., bulk, emulsion, and suspension. However, continuous bulk polymerization in well mixed continuous stirred tank reactors (CSTR) is the preferred process for the manufacture of SAN resins for applications requiring high optical quality. The superior color and haze of SAN made by bulk polymerization is due to the lack of emulsifiers, suspending agents, salt, and water found in the alternate processes. Bulk processes of the CSTR type [1, 2] are required to eliminate copolymer composition drift. It has been determined that as little as a 4% drift in composition leads to haze in SAN [3]. As with polystyrene, many of the uses for SAN require high optical quality. However, the introduction of AN monomer units into the polystyrene backbone has a very detrimental effect upon color stability.

In comparison to polystyrene, developing an understanding of SAN degradation chemistry is very challenging. The complexity of SAN is due to the many variables that the introduction of a second monomer unit brings. Examples of copolymerization variables that have been linked to SAN optical quality include: AN content [4, 5], AN sequence distribution in the polymer [6–8], oligomer structures [9, 10], impurities in the AN monomer [11], monomer side reactions during copolymerization [12], the potential for copolymer composition drift [3, 13], dissolved oxygen in polymerization feed [14], and initiator type [15, 16]. All of these factors likely influence the optical quality of SAN and must be considered before high quality SAN can be produced. However, compared to polyacrylonitrile (PAN), detailed studies aimed at elucidation of mechanisms involved in the thermal discoloration of SAN are few [6, 17]. This lack of attention could be partly due to the assumption that the main mechanism of discoloration of SAN involves cyclization of AN sequences in the SAN backbone to form conjugated polyimines (Scheme 1) [6].

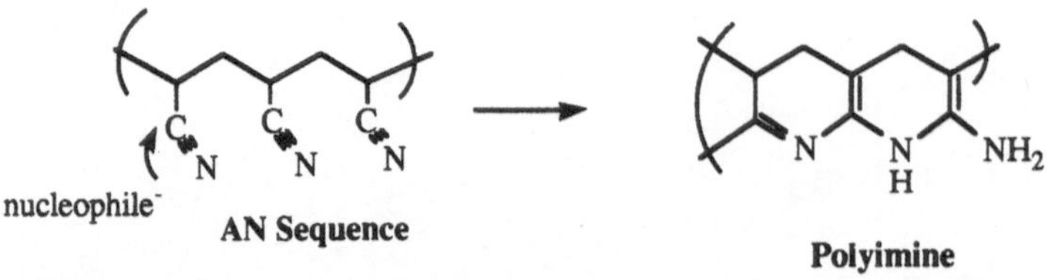

Scheme 1. Cyclization of an AN sequence to form a conjugated polyimine postulated as a possible mechanism of discoloration of AN containing polymers [6, 18]

Any attempt to understand the chemistry of thermal discoloration of SAN must begin with an understanding of the degradation chemistry of polyacrylonitrile (PAN). Relative to SAN, PS is very discoloration resistant. PAN, on the other hand, discolors rapidly upon heating. Thus, it is often assumed that the discoloration chemistry of SAN involves AN sequences in the SAN backbone. Therefore, an understanding of PAN discoloration chemistry is important. There have been several excellent reviews of PAN discoloration chemistry [19–22] and therefore it will not be reviewed here.

2 The Chemistry of the Polymer Backbone

2.1 The Mechanism of Copolymerization

The development of an understanding of SAN backbone microstructure first requires a discussion of the controlling factors. The microstructure of the SAN backbone is controlled by the kinetics of the free radical chain growth process, rather than by thermodynamics. The copolymerization kinetics of SAN has been the subject of numerous studies over the past fifty years. In most studies, the terminal copolymerization model, originally developed in 1944 by the independent studies of Alfrey and Goldfinger, Mayo and Lewis, and Wall, has been applied to SAN copolymerization. The terminal model assumes that copolymer composition is controlled only by the identity of the terminal unit on the growing chain. In this model, copolymer composition is controlled by four propagation steps (Scheme 2) and many reactivity ratio values for SAN copolymerization (r_A, r_S) have been calculated according to this model (Table 1).

Terminal Model

$$\rightsquigarrow S\cdot \; + S \xrightarrow[k_{SS}]{} \rightsquigarrow SS\cdot$$

$$\rightsquigarrow S\cdot \; + A \xrightarrow[k_{SA}]{} \rightsquigarrow SA\cdot$$

$$\rightsquigarrow A\cdot \; + S \xrightarrow[k_{AS}]{} \rightsquigarrow AS\cdot$$

$$\rightsquigarrow A\cdot \; + A \xrightarrow[k_{AA}]{} \rightsquigarrow AA\cdot$$

$r_S = k_{SS}/k_{SA}$ $\qquad$ $r_A = k_{AA}/k_{AS}$ $\qquad$ **Scheme 2.** The terminal model for copolymerization

Table 1. Literature terminal reactivity ratio values for SAN copolymerization

r_A	r_S	$r_A r_S$	Temp. (°C)	Reference
0.011	0.29	0.003	− 30	[23]
0.02	0.34	0.007	5	[23]
0.05	0.38	0.019	41.5	[24]
0.05	0.37	0.019	50	[25]
0.05	0.37	0.026	50	[25]
0.10	0.27	0.027	50	[26]
0.04	0.41	0.016	50	[27]
0.04	0.41	0.016	60	[28]
0.04	0.40	0.016	60	[29]
0.13	0.25	0.033	60	[30]
0.16	0.30	0.048	60	[30]
0.20	0.40	0.080	60	[30]
0.16	0.30	0.048	60	[31]
0.13	0.36	0.047	60	[31]
0.13	0.40	0.052	60	[31]
0.12	0.32	0.038	60	[32]
0.12	0.32	0.038	60	[33]
0.17	0.30	0.051	60	[34]
0.15	0.33	0.050	60	[34]
0.07	0.46	0.032	60	[34]
0.03	0.43	0.013	60	[34]
0.02	0.45	0.009	65	[24]
0.04	0.41	0.016	65	[23]
0.04	0.41	0.016	70	[19]
0.03	0.41	0.012	75	[35]
0.17	0.33	0.056	80	[34]
0.17	0.36	0.061	80	[34]
0.06	0.41	0.025	80	[34]
0.06	0.55	0.033	80	[34]
0.02	0.47	0.009	86.5	[24]
0.06	0.39	0.023	99	[36]
0.067	0.47	0.031	130	[23]

This model leads to a prediction of copolymer sequence distribution based upon Markovian statistics. Following statistical derivation of the terminal copolymerization model, the mole fraction of any sequence can be expressed as a combination of the conditional probabilities for adding a certain monomer, given that the growing chain ends in a certain reactive group.

$p(A|S)$ = probability of adding A, given that the growing chain ends with a S terminal group

$p(S|S) = 1 - p(A|S)$

$p(S|A)$ = probability of adding S, given that the growing chain ends with an A terminal group

$p(A|A) = 1 - p(S|A)$

The conditional probabilities expressed in terms of the reactivity ratios, r_A and r_S, and the monomer feed ratio, F.

$$F = \frac{[A]}{[S]}$$

$$p(A|S) = \left(\frac{F}{F + r_S}\right) \quad p(S|A) = \left(\frac{1}{1 + Fr_A}\right)$$

The triad sequence distributions can be expressed in terms of the conditional probabilities in Table 2.

Theoretically, if the terminal reactivity ratio data from the literature is of sufficient quality, Arrhenius plots would show the temperature sensitivity of the reactivity ratios and one could calculate activation energy differences for the addition of S and AN monomers to S and AN terminal growing polymer radicals, respectively (Scheme 3) [13, 24, 37].

$$r_A = k_{AA}|k_{AS}; r_S = k_{SS}|k_{SA}$$

$$k_{AA} = A_{AA}e^{-Ea_{AA}/RT} \quad k_{AS} = A_{AS}e^{-Ea_{AS}/RT}$$

Table 2. Expressions for composition and triad fractions in terms of first-order Markov conditional probabilities

Sequence	Sequence mole fraction
A (mole fraction)	$\dfrac{p(A|S)}{p(A|S) + p(S|A)}$
AAA	$\dfrac{p(A|S)}{p(A|S) + p(S|A)} \times p(A|A)^2$
AAS, SAA	$2 \times \dfrac{p(A|S)}{p(A|S) + p(S|A)} \times p(A|A) \times p(S|A)$
ASA	$\dfrac{p(A|S)}{p(A|S) + p(S|A)} \times p(S|A) \times p(A|S)$
ASS, SSA	$2 \times \dfrac{p(A|S)}{p(A|S) + p(S|A)} \times p(S|A) \times p(S|S)$
SAS	$\dfrac{p(S|A)}{p(A|S) + p(S|A)} \times p(A|S) \times p(S|A)$
SSS	$\dfrac{p(S|A)}{p(A|S) + p(S|A)} \times p(S|S)^2$

$$r_A = \frac{A_{AA}}{A_{AS}} e^{-(Ea_{AA} - Ea_{AS})/RT}$$

$$r_A = e^{-(Ea_{AA} - Ea_{AS})/RT} \longrightarrow 1 \text{ as } T \to \infty$$

$$r_A < 1 \longrightarrow Ea_{AA} > Ea_{AS}$$

$$r_A > 1 \longrightarrow Ea_{AA} < Ea_{AS}$$

$$\ln r_A = -[(Ea_{AA} - Ea_{AS})/R]\, 1/T$$

$$\text{Slope} < 0 \longrightarrow Ea_{AA} > Ea_{AS}$$

$$\text{Slope} > 0 \longrightarrow Ea_{AA} < Ea_{AS}$$

Scheme 3. Relationship of reactivity ratios to Arrhenius parameters

A plot of all of the terminal reactivity ratio data from the literature (Table 1) shows, even with the significant scatter (especially the r_A numbers), that r_A and r_S approach unity at infinite temperature (Fig. 1). The data scatter is likely due to the large number of researchers using a variety of techniques for reactivity ratio calculation. Plotting the data of one researcher alone should solve this problem. Johnston [23] was chosen because he collected reactivity ratio data for bulk SAN copolymerization over a broad temperature range (−30–130 °C). Indeed,

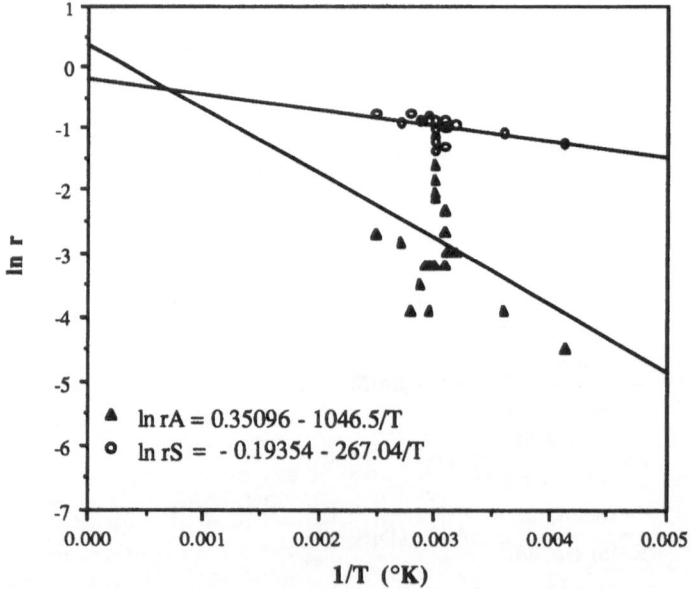

▲ $\ln r_A = 0.35096 - 1046.5/T$
○ $\ln r_S = -0.19354 - 267.04/T$

Fig. 1. Arrhenius plot (least squares fit) of the literature terminal reactivity ratio data for SAN copolymerization versus temperature

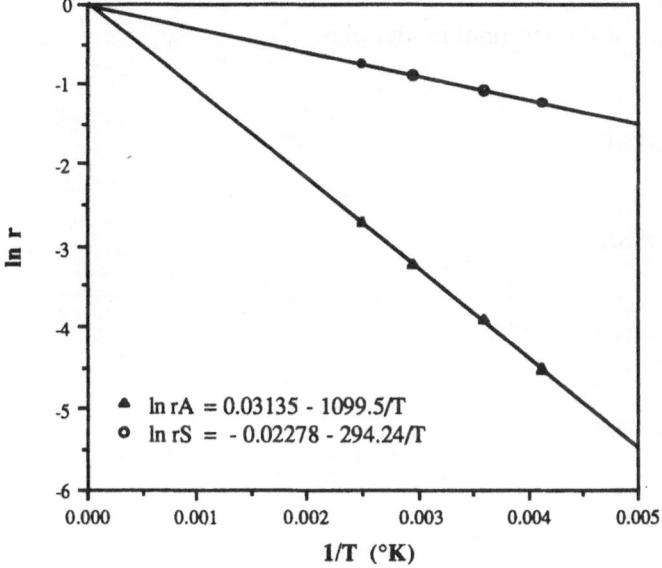

Fig. 2. Arrhenius plot of Johnston's terminal reactivity ratio data for SAN copolymerization versus temperature [23]

the Arrhenius plot of his data almost perfectly intersects at the ln of unity (Fig. 2).

2.2 Measurement of Sequence Distribution and Alternate Copolymerization Models

Many analytical techniques have been utilized to analyze the SAN microstructure including: LALLS [38], ^{13}C NMR [19, 31, 39–44], infrared spectroscopy [45–49], ultraviolet spectroscopy [50–52], pyrolysis GC [8, 27, 53], pyrolysis mass spectroscopy [54, 55], fluorescence [20, 56], GPC-IR [57, 58], and GPC-UV [52]. Since the terminal model allows the calculation of sequence distribution, the calculated and measured sequence distributions can be compared. This comparison generally shows deviation of the measured sequence distribution vs that predicted using the terminal model. Ham [59] was the first to notice the deviation and explained the deviation based upon penultimate effects. Since that time several other researchers have also noticed deviation of their data from the terminal model and have applied more elaborate copolymerization models (Scheme 4) to explain the mechanism of SAN copolymerization. The penultimate [60, 61] and complex participation models [33, 62, 63] have both been evaluated and give a better fit to the SAN system than the terminal model.

Complex Participation model

four propagation steps of the terminal model plus:

$$S + A \rightleftharpoons SA$$

$$\leadsto S\cdot + \overline{AS} \xrightarrow[k_{S\overline{AS}}]{} \leadsto SAS\cdot$$

$$\leadsto S\cdot + \overline{SA} \xrightarrow[k_{S\overline{SA}}]{} \leadsto SSA\cdot$$

$$\leadsto A\cdot + \overline{AS} \xrightarrow[k_{A\overline{AS}}]{} \leadsto AAS\cdot$$

$$\leadsto A\cdot + \overline{SA} \xrightarrow[k_{A\overline{SA}}]{} \leadsto ASA\cdot$$

$$r_S = k_{SS}/k_{SA} \qquad r_A = k_{AA}/k_{AS}$$

$$p_S = k_{S\overline{SA}}/k_{S\overline{AS}} \qquad p_A = k_{A\overline{AS}}/k_{A\overline{SA}}$$

$$s_S = k_{S\overline{AS}}/k_{SA} \qquad s_A = k_{A\overline{SA}}/k_{AS}$$

Penultimate Model

$$\leadsto SS\cdot + S \xrightarrow[k_{SSS}]{} \leadsto SSS\cdot$$

$$\leadsto SS\cdot + A \xrightarrow[k_{SSA}]{} \leadsto SSA\cdot$$

$$\leadsto AS\cdot + S \xrightarrow[k_{ASS}]{} \leadsto ASS\cdot$$

$$\leadsto AS\cdot + A \xrightarrow[k_{ASA}]{} \leadsto ASA\cdot$$

$$\leadsto SA\cdot + A \xrightarrow[k_{SAA}]{} \leadsto SAA\cdot$$

$$\leadsto SA\cdot + S \xrightarrow[k_{SAS}]{} \leadsto SAS\cdot$$

$$\leadsto AA\cdot + A \xrightarrow[k_{AAA}]{} \leadsto AAA\cdot$$

$$\leadsto AA\cdot + S \xrightarrow[k_{AAS}]{} \leadsto AAS\cdot$$

$$r_{SS} = k_{SSS}/k_{SSA} \qquad r_{AA} = k_{AAA}/k_{AAS}$$

$$r_{AS} = k_{ASS}/k_{ASA} \qquad r_{SA} = k_{SAA}/k_{SAS}$$

Scheme 4. Models used to fit SAN sequence distribution data

Explanations, other than the inadequacy of the terminal model, have been given to explain potential causes for deviation. Pichot et al. [30] offered several possible explanations for these discrepancies including: 1) preferential solvation of one of the monomers in the polymer; 2) AN existing as a dimer due to dipole–dipole interactions; and 3) terminal radical interaction with the AN nitrile group. Harwood [64] presents evidence indicating that it is the monomer concentrations local to the active radical center that controls the copolymerization and backbone monomer sequence distribution rather than the average monomer concentrations in the reactor. Harwood calls this the "bootstrap model" because it is the nature of the polymer chain itself that controls the local monomer concentration near its active chain-end.

The use of sequence distribution measurements to compare calculated with actual sequence data is a common technique to evaluate the accuracy of copolymerization models. Recently, Hill et al. [40, 56, 65, 66] compared all three models and found that the penultimate model gave the best fit to their NMR sequence distribution data obtained from SAN copolymers made using bulk polymerization at 60 °C.

The ^{13}C NMR triad data from four different research groups is shown in Tables 3–6. The triad data shown in these tables has been recalculated. In the original references, the six triads are separated into two groups, i.e., the three S centered triads and the three A centered triads, and each group is normalized to 1. In this review, the three S triads have been normalized to add up to the

Table 3. Hill et al. [65] NMR triad distribution for SAN copolymerization in toluene at 60 °C

A	SSS	SSA + AAS	ASA	AAA	AAS + SAA	SAS
0.869		0.004	0.127	0.578	0.246	0.045
0.875		0.006	0.119	0.619	0.228	0.029
0.722		0.022	0.256	0.263	0.387	0.072
0.633		0.040	0.327	0.091	0.338	0.204
0.558	0.007	0.099	0.336	0.045	0.234	0.279
0.562	0.013	0.105	0.320	0.039	0.242	0.281
0.515	0.011	0.146	0.329	0.018	0.184	0.313
0.484	0.012	0.199	0.305	0.010	0.137	0.336
0.451	0.031	0.254	0.264	0.004	0.091	0.356
0.402	0.068	0.329	0.200	0.004	0.053	0.344
0.356	0.124	0.370	0.150		0.031	0.325
0.300	0.208	0.395	0.097		0.016	0.284
0.223	0.385	0.341	0.051			0.223
0.215	0.371	0.374	0.039		0.007	0.208

Table 4. Pichot et al. [13] NMR sequence distribution data for SAN copolymerization in toluene at 60 °C

A	SSS	SSA + ASS	ASA	AAA	AAS + SSA	SAS
0.930				0.725	0.205	
0.700			0.300	0.333	0.315	0.060
0.640				0.141	0.358	0.141
0.540				0.027	0.189	0.324
0.350		0.319	0.345		0.056	0.294
0.270	0.256		0.090			0.270
0.090	0.683	0.228				0.090

Table 5. Allan et al. [67] NMR sequence distribution data for SAN copolymerization in ethylbenzene in a CSTR process at 150–160 °C

A	SSS	SSA + ASS	ASA	AAA	AAS + SAA	SAS
0.250	0.330	0.320	0.070		0.020	0.260
0.340	0.170	0.370	0.120		0.040	0.300
0.390	0.110	0.340	0.160		0.070	0.320
0.470	0.050	0.250	0.230	0.010	0.160	0.300
0.520	0.020	0.190	0.270	0.030	0.210	0.280
0.580	0.010	0.100	0.300	0.060	0.300	0.230

Table 6. Arita et al. [19] NMR sequence distribution data for SAN copolymerization in bulk at 70 °C

A	SSS	SSA + ASS	ASA	AAA	AAS + SAA	SAS
0.690			0.310	0.145	0.407	0.138
0.620		0.042	0.338	0.087	0.329	0.205
0.500		0.200	0.300		0.200	0.300
0.490		0.199	0.311		0.167	0.323
0.370	0.113	0.328	0.189		0.044	0.326
0.250	0.338	0.345	0.068			0.250

mole fraction of S in the copolymer and the three A centered triads normalized to add up to the mole fraction of AN in the copolymer. Graphs of the triad data (Figs. 3–8) show fairly good agreement. The data of Allan et al. [67] shows the greatest deviation from the average triad sequence data reported by the other three research groups. This deviation is likely due to the source of the samples analyzed. Allan et al. analyzed SAN samples that had been prepared using a continuous commercial CSTR bulk polymerization process at high monomer conversion (> 50%) and temperature (150–160 °C) while the other three groups prepared their copolymers using low conversion (< 10%) batch polymerization at low temperatures (60 °C). If the product of the reactivity ratios are calculated from the sequence distribution data, the data of Allan et al. leads to an average value significantly higher than previously reported in the literature. This indicates that SAN prepared in continuous commercial bulk polymerization is less

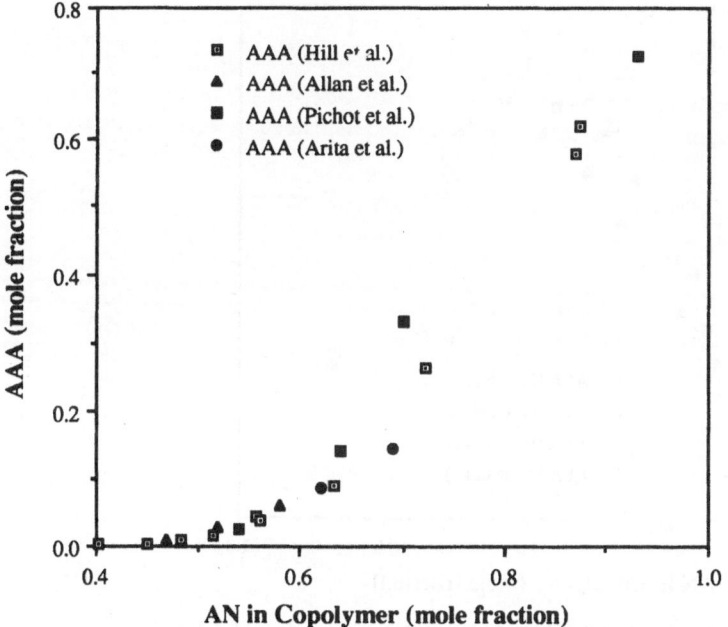

Fig. 3. Literature values for AAA triads in SAN vs AN content of copolymer

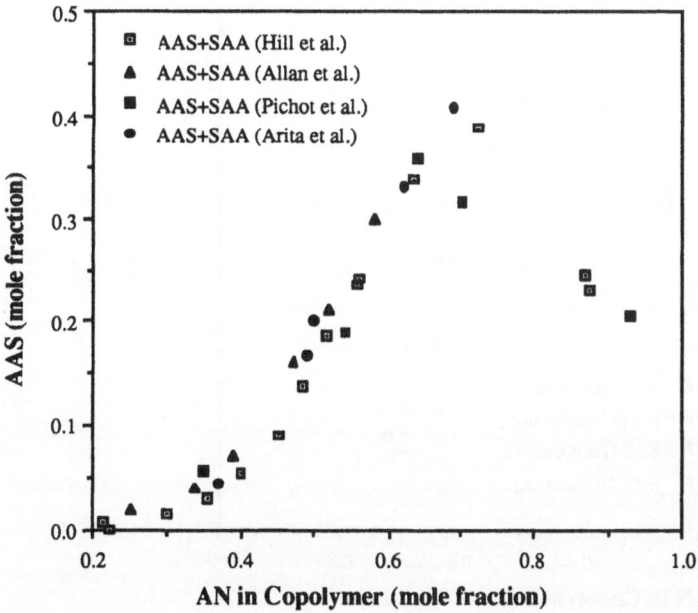

Fig. 4. Literature values for AAS +SAA triads in SAN vs AN content of copolymer

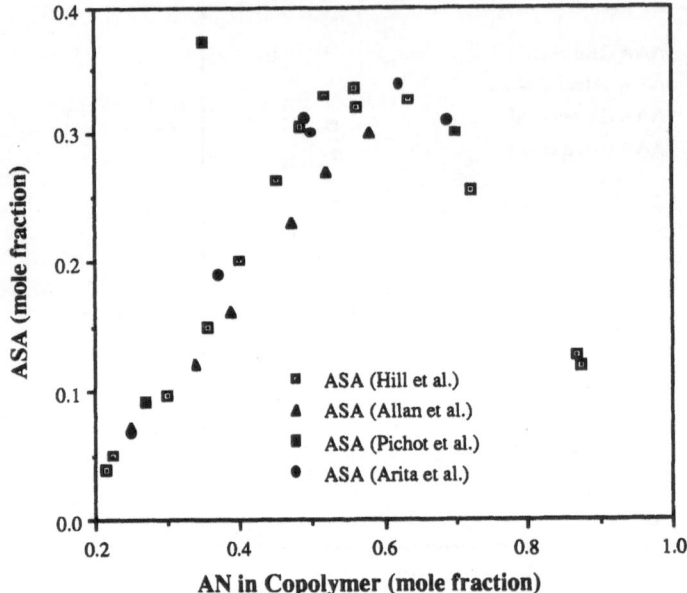

Fig. 5. Literature values for ASA triads in SAN vs AN content of copolymer

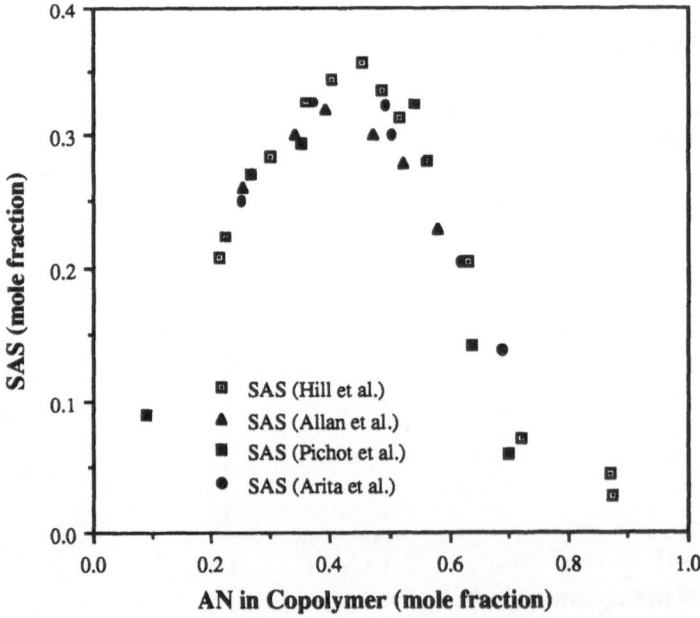

Fig. 6. Literature values for SAS triads in SAN vs AN content of copolymer

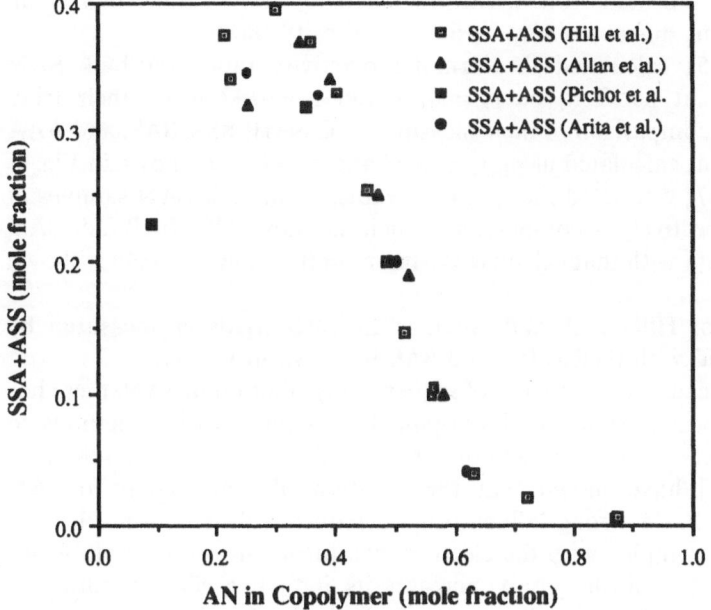

Fig. 7. Literature values for SSA +ASS triads in SAN vs AN content of copolymer

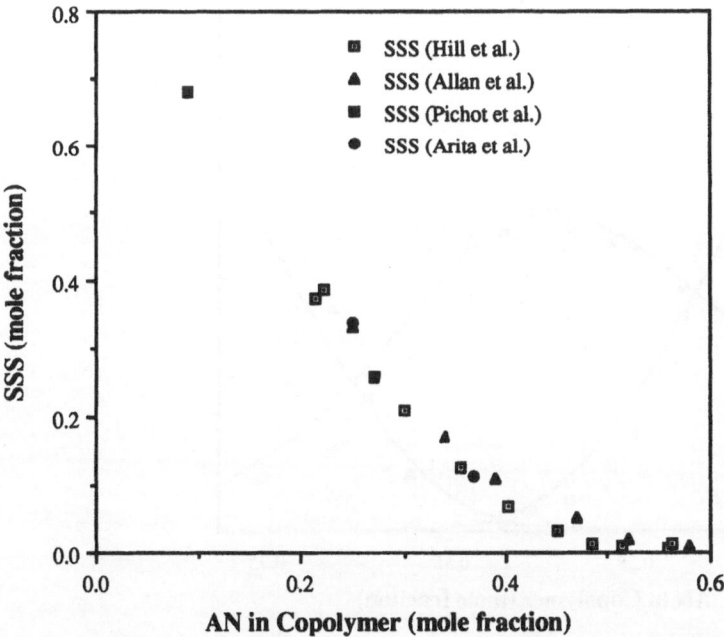

Fig. 8. Literature values for SSS triads in SAN vs AN content of copolymer

alternating than indicated from literature reactivity ratio data. Allan et al. postulated that this difference may be due to viscosity effects.

Hill et al. [65] calculated the terminal reactivity ratios for bulk SAN copolymerization at 60 °C to be $r_A = 0.08$ and $r_S = 0.47$ using their triad sequence data. A comparison of their measured ^{13}C NMR SSS, SAS, and AAA triad data with that calculated using $r_A = 0.08$ and $r_S = 0.47$ is shown in Fig. 9.

Allan et al. [67] estimated the r_A and r_S values from their SAN samples at 0.18 and 0.49, respectively. A comparison of their measured ^{13}C NMR SSS, SAS, and AAA triad data with that calculated using $r_A = 0.18$ and $r_S = 0.49$ is shown in Fig. 10.

Comparison of Hill et al. and Allan et al. SAS triads as measured by ^{13}C NMR overlaid with their calculated SAS is shown in Fig. 11.

Whatever the cause for deviation of sequence distribution in SAN, there has been considerable interest in the development of polymerization methods to minimize the AN sequence length [6, 7, 58, 68–75]. Many researchers [6, 7, 71–73, 75, 76] have shown that the addition of Lewis acids to SAN copolymerization has a strong influence upon sequence distribution. Electron poor Lewis acids complex with the electron rich nitrile group of acrylonitrile. Complexation of the nitrile group increases its inductive effect resulting in a decrease of electron density of the acrylonitrile double bond (its Alfrey–Price e-value becomes more positive). Thus copolymerization of the acrylonitrile:

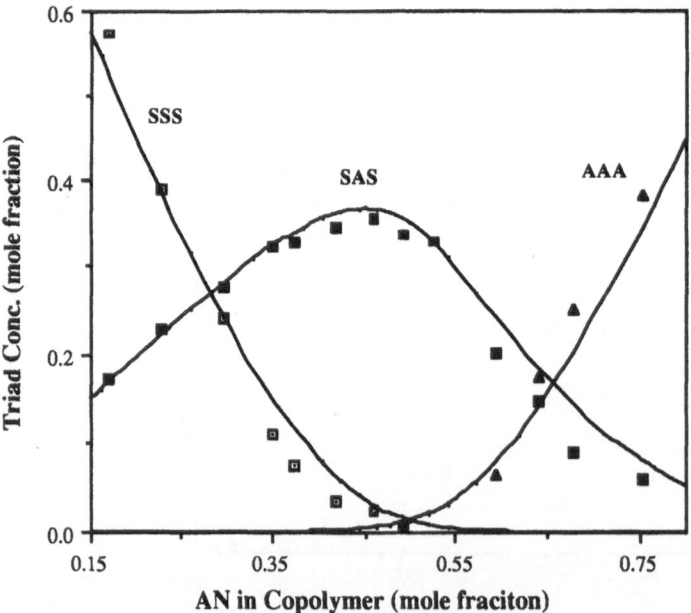

Fig. 9. Comparison of the measured (^{13}C NMR) triad data (Hill et al. [65]) with that calculated using $r_A = 0.08$ and $r_S = 0.47$

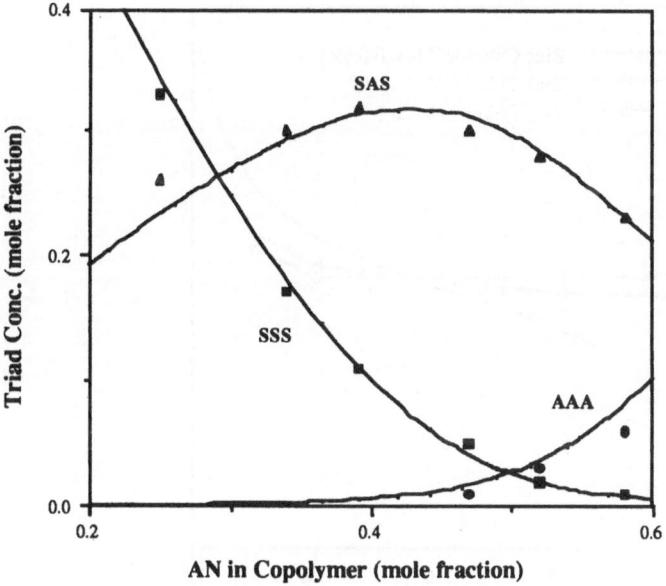

Fig. 10. Overlay of the actual AN triads with the triads calculated using $r_A = 0.18$ and $r_S = 0.49$

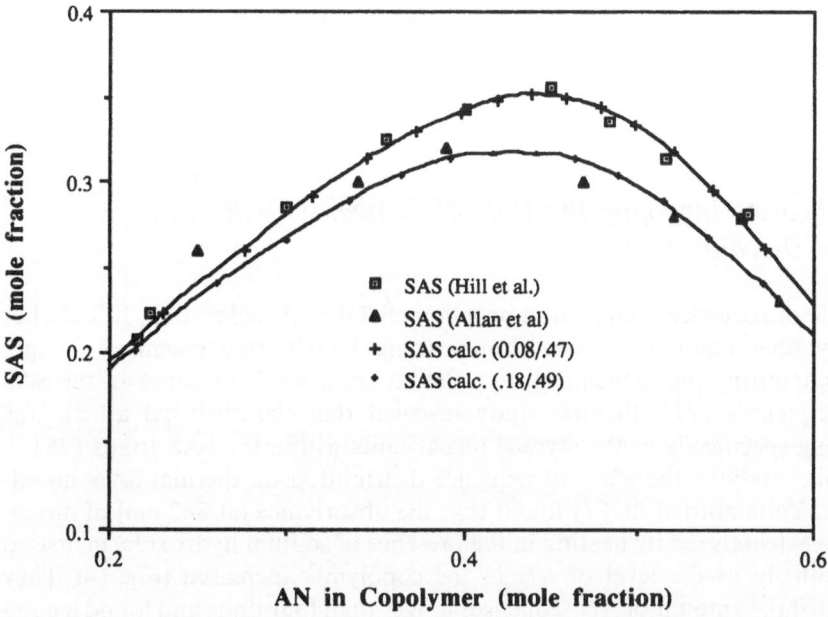

Fig. 11. Overlay of the actual AN triads with the triads calculated usng $r_A = 0.08$ and $r_S = 0.47$

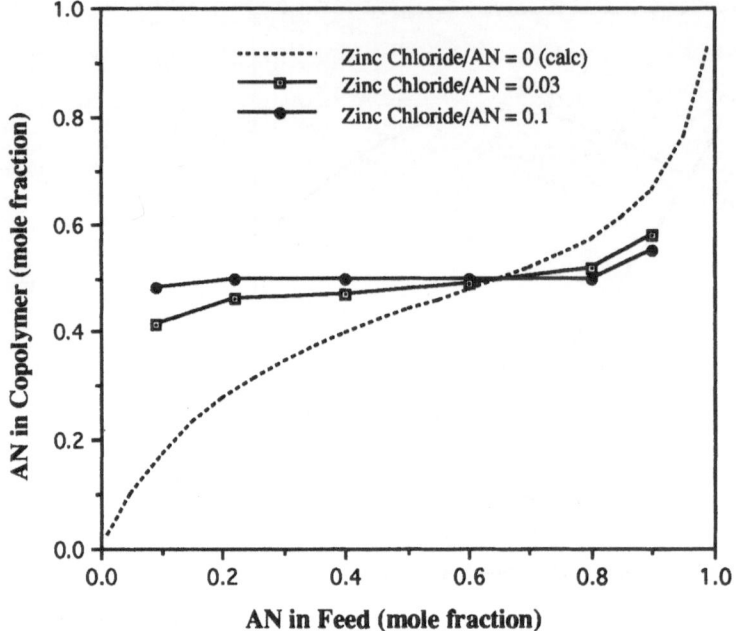

Fig. 12. Comparison of the SAN copolymer curves obtained at three levels of zinc chloride [6]

Lewis acid complex with styrene results in the formation of styrene-alt-acrylo-nitrile having far superior color stability to styrene-stat-acrylonitrile. Compari-son of the SAN copolymer curves obtained at three levels of zinc chloride is shown in Fig. 12 [6].

Both r_A and r_S decrease as the level of zinc chloride is increased as shown in Fig. 13.

2.3 Theories Involving the Role of Sequence Distribution on Discoloration

The role of sequence distribution on photochemical discoloration of SAN has recently been reported. It was shown, using FT-IR, that chemical changes occurring during photochemical degradation are a result of attack of the SAS triad sequences [77]. Further study revealed that the chemical attack was occurring specifically at the styrene repeat units within the SAS triads [78].

While studying the effect of sequence distribution on thermal SAN discol-oration, Yubamoto et al. [7] found that the absorbance (at 362 nm) of discol-ored SAN (catalyzed by heating in the presence of sodium hydroxide) increased exponentially as the level of AN in the copolymer increased (Fig. 14). They calculated the amount of AN sequences longer than four units and found a good correlation (Fig. 15).

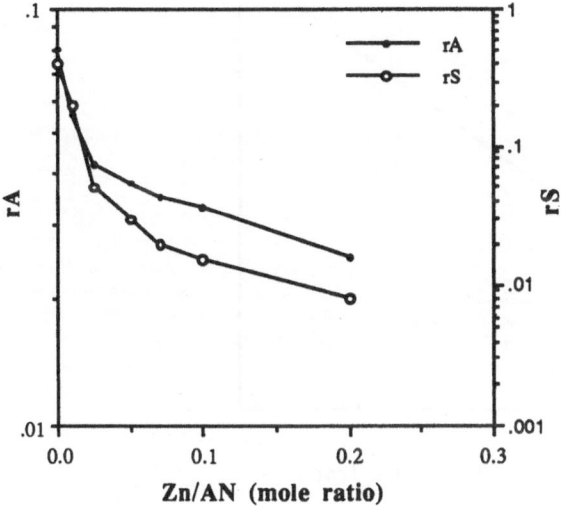

Fig. 13. Effect of ZnCl$_2$ level on reactivity ratios [6]

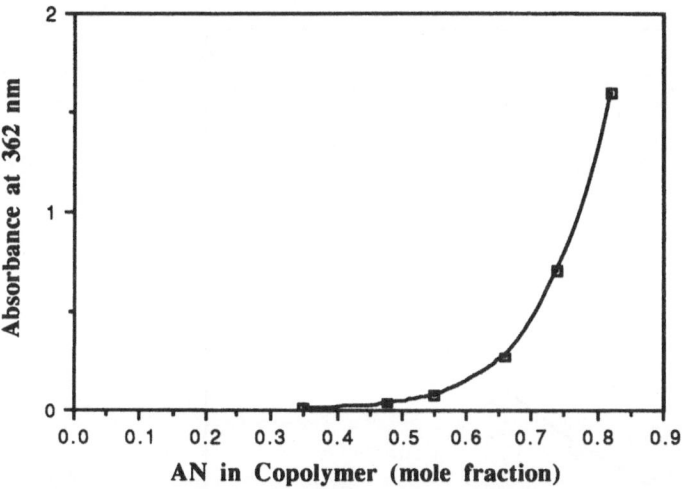

Fig. 14. Increase in absorbance of SAN after heating in the presence of NaOH as a function of AN content [7]

Allan et al. [67] also observed a correlation of SAN discoloration with AN sequences. They molded a series of SAN copolymers of varying AN content and observed an exponential increase of color with both temperature (Fig. 16) and AN content (Fig. 17).

If the discoloration chemistry is due mainly to the cyclization of AN sequences, the increase in discoloration with increasing AN content should correspond to the increase in AN sequences. The plot of discoloration during molding at 280 °C vs AN triad concentration in the copolymers as measured by

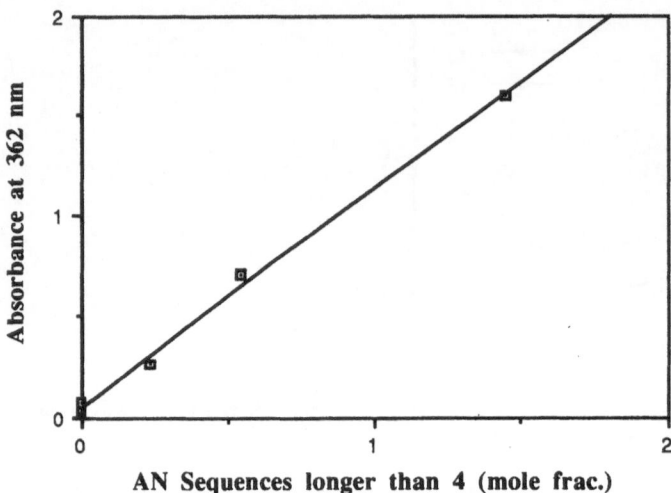

Fig. 15. Absorbance of SAN after heating in the presence of NaOH as a function of AN sequences longer than 4 [7]

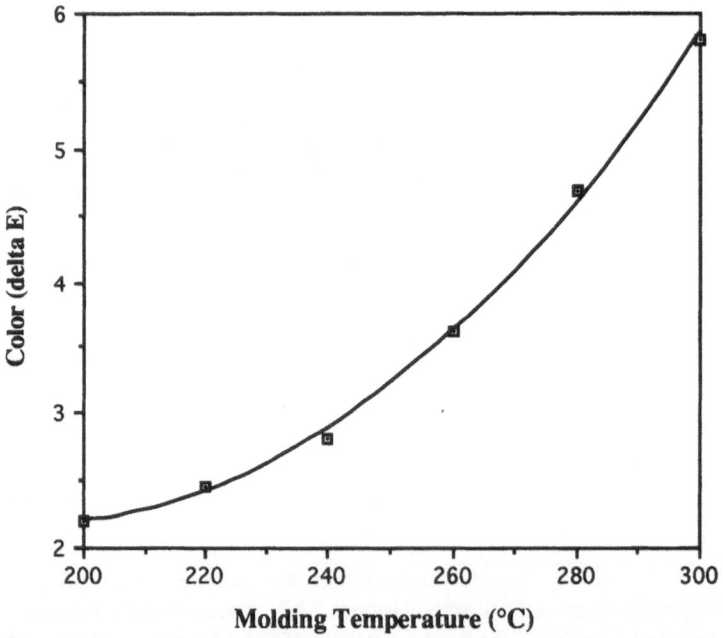

Fig. 16. Increase in the color (ΔE) of SAN (0.433 mole fraction AN) with molding temperature [67]

^{13}C NMR (Fig. 18) shows some correlation suggesting that AN cyclization chemistry may be a contributor to SAN discoloration during molding.

To separate small molecule chromophores from the chromophores in the polymer backbone, Allan et al. [67] utilized GPC-UV/vis [79–81]. Five bulk

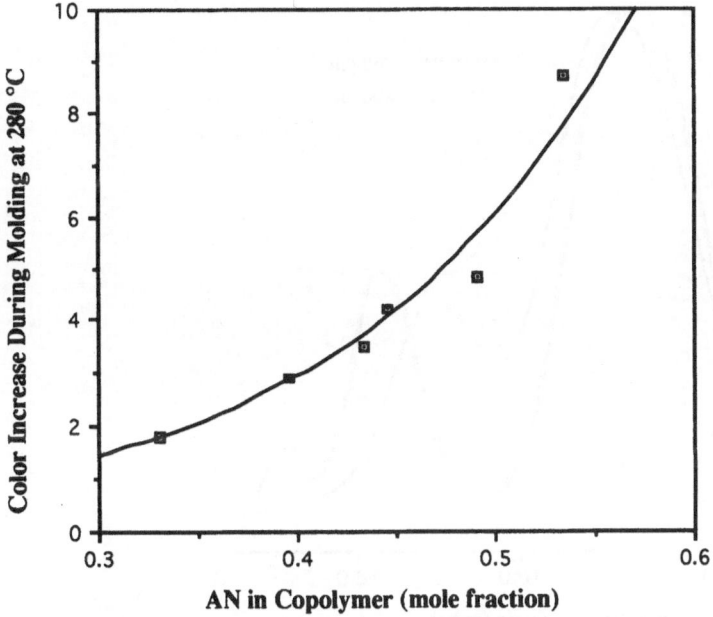

Fig. 17. Increase in SAN color during molding at 280 °C as AN content of SAN increases [67]

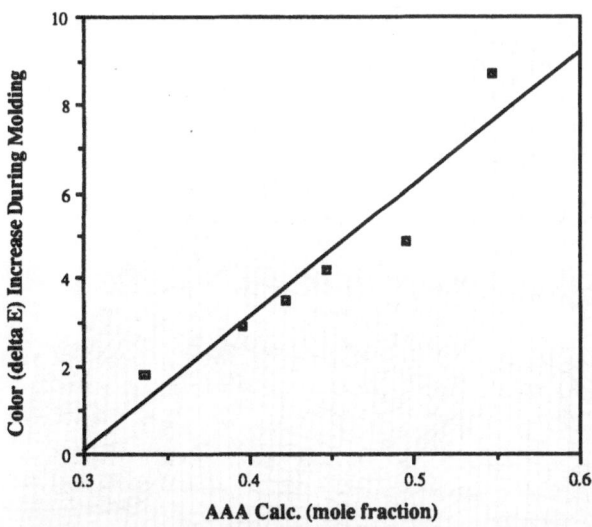

Fig. 18. Color increase during molding at 280 °C vs the AN triads measured by ^{13}C NMR [67]

CSTR prepared SAN copolymers having different AN contents were thermolyzed in evacuated (<4 mm Hg) glass ampules at 250 °C for 72 h and analyzed using GPC-UV/vis. Figure 19 shows an overlay of GPC curves collected simultaneously at 260 and 400 nm of the thermolyzed SAN resin (0.492 mole

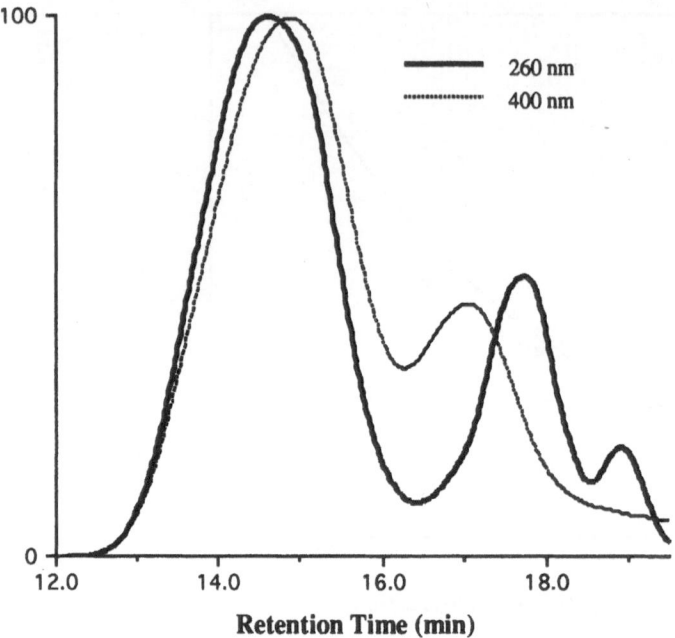

Fig. 19. Overlay of the 260 nm and 400 nm GPC curves [67]

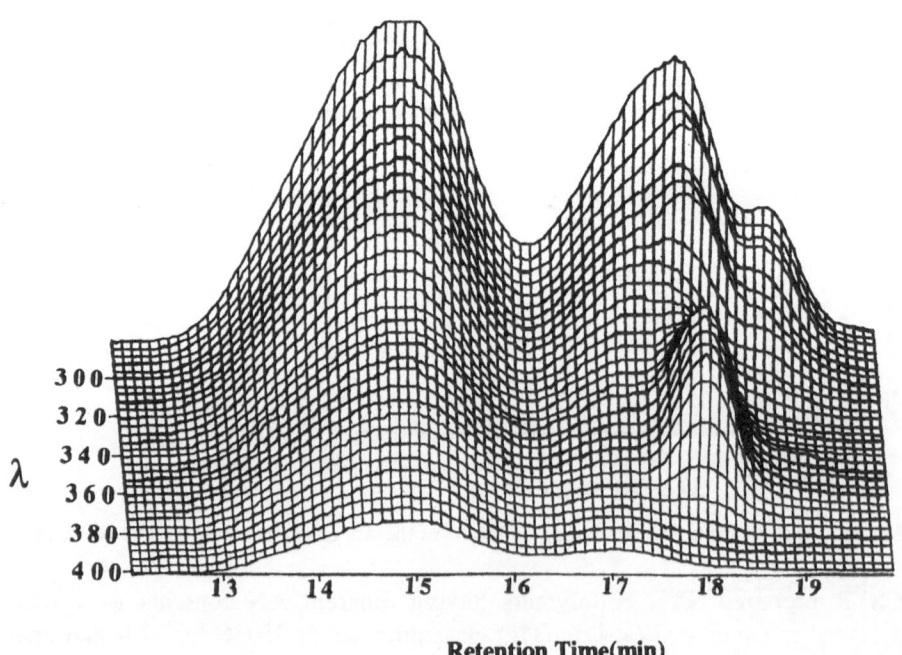

Fig. 20. 3-D view of GPC analysis of discolored SAN [67]

fraction AN). Figure 20 shows the 3-D plot (Absorbance vs Wavelength vs Time) of the same sample. Both the overlay of the 260 and 400 nm GPC curves (Fig. 19) and the 3-D view of the GPC (Fig. 20) show that absorbance in the visible is present in both high and low polymer.

The area under the high polymer portion of the 400 nm GPC curves increased with AN content (Fig. 21). Much improved correlation was found between the area under the 400 nm GPC curve of the high polymer portion of five samples vs the calculated AN triad content of the copolymers (Fig. 22) when compared to the correlation achieved by measuring bulk discoloration (Fig. 18).

The chromophore in the high polymer portion appears to be located primarily on the chain-ends as evidenced by an offset of the 400 nm GPC curve to the low molecular weight side of the 260 nm GPC curve (Fig. 19) [79–81]. The offset occurs because the chromophore causing the 400 nm absorbance increases in concentration as the chain length decreases while the phenyl rings giving rise to the 260 nm signal are in equal concentration in all of the polymer chains, independent of their molecular weight. Grassie et al. [82, 83], proposed the formation of a chain-end chromophore during SAN degradation. While studying the thermal degradation of acrylonitrile copolymers, they found that styrene units were a strong blocking agent for nitrile cyclization. Originally, they believed that the blocking mechanism was purely steric [83] but later, they changed their opinion [82]. Since it was previously shown [84] that chain scission occurs quite readily at acrylonitrile–styrene bonds, they believe that the nitrile radical intermediate formed during nitrile cyclization, transfers to a styrene unit during the chain breaking process (Scheme 5). The terminal styryl

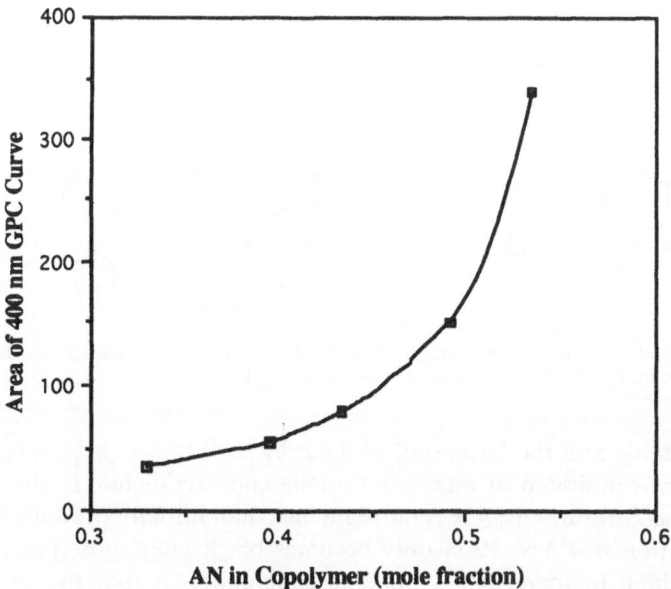

Fig. 21. Area under high polymer portion of 400 nm GPC curve vs AN content [67]

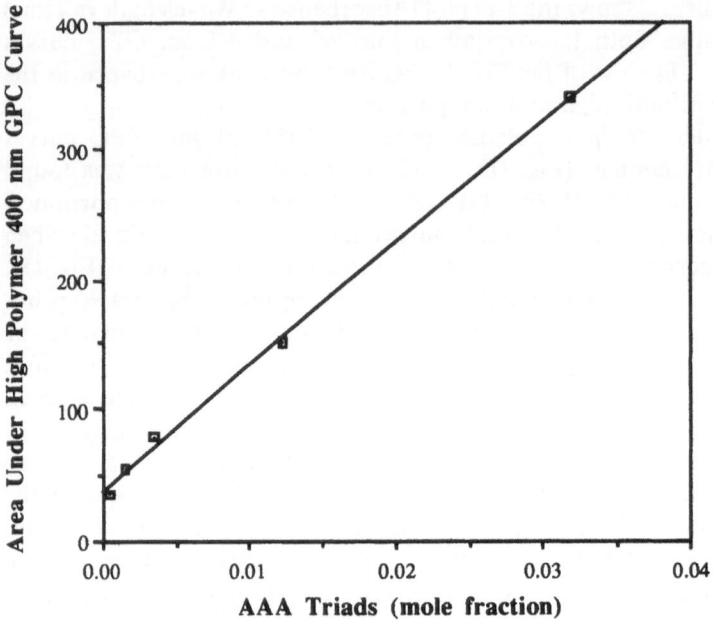

Fig. 22. Area of high polymer portion of 400 nm GPC curve of discolored SAN vs AAA triad concentration calculated using $r_A = 0.18$ and $r_S = 0.49$ [67]

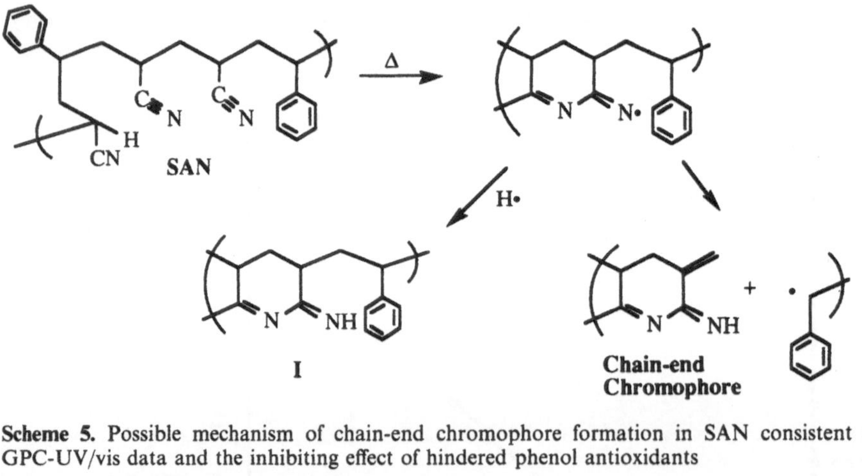

Scheme 5. Possible mechanism of chain-end chromophore formation in SAN consistent with GPC-UV/vis data and the inhibiting effect of hindered phenol antioxidants

radical is relatively stable and the formation of the polymer radicals necessary for further cyclization is inhibited or retarded. Grassie and McGuchan further point out that the discoloration of SAN is more intense than initially produced during nitrile cyclization in PAN. PAN only becomes black after subsequent dehydrogenation at high temperature. A possible explanation is that the enhanced discoloration of SAN is caused by overlap of the cyclization and dehydrogenation reactions. However, they found by Thermal Volatilization Analysis (TVA) that the amount of hydrogen evolved during thermal decomposition of SAN is extremely small. An alternative explanation they offered was that the increased chain scission in SAN (Scheme 5) produces quinoid structures by dehydrogenation of the thermal heterocycle which would greatly increase the chromophoric activity. The improvement in color stability by the addition of hindered phenol antioxidants [85] also supports the Grassie mechanism since the hindered phenol would quench the terminal iminyl radical (I) before chain scission and subsequent chromophore formation takes place (Scheme 5).

3 Inhibition of Backbone Discoloration

Runge and Nelles noticed that the addition of maleic anhydride to nitrile containing polymers stabilizes them against thermal discoloration [86]. The mechanism of stabilization they propose involves the interception of the intermediate imine nucleophile formed after initiation of nitrile cyclization within AN sequences, thus terminating the nitrile cyclization process (Scheme 6).

Marien, however, discounts this mechanism because he found that succinic anhydride makes the discoloration worse [17]. He therefore believes that maleic anhydride is acting as a Diels–Alder dienophile rather than an electrophile. Furthermore, Marien found that most strong dienophiles improve the discoloration resistance of SAN (Fig. 23).

For this mechanism to be operative, the incipient discoloration intermediate must be a diene capable of undergoing a Diels–Alder reaction. Marien rules out the nitrile cyclization mechanism proposed by Johnson et al. [18] (Scheme 1) since he found that a model of the naphthpyridine ring system was unreactive toward maleic anhydride (Scheme 7).

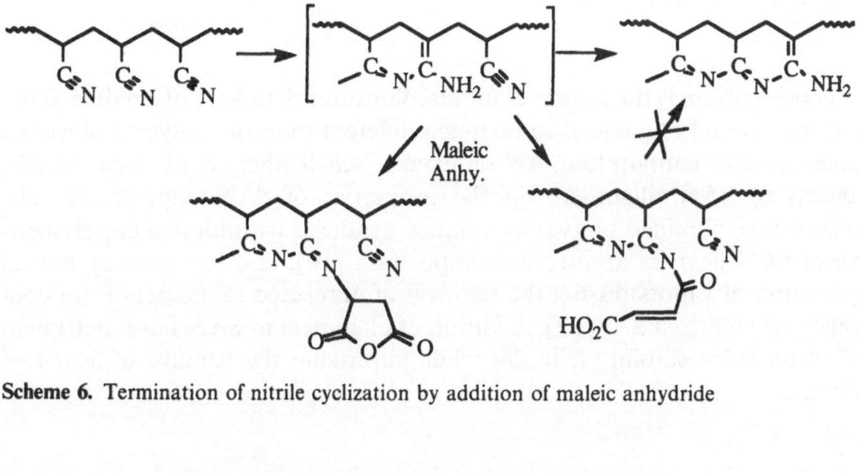

Scheme 6. Termination of nitrile cyclization by addition of maleic anhydride

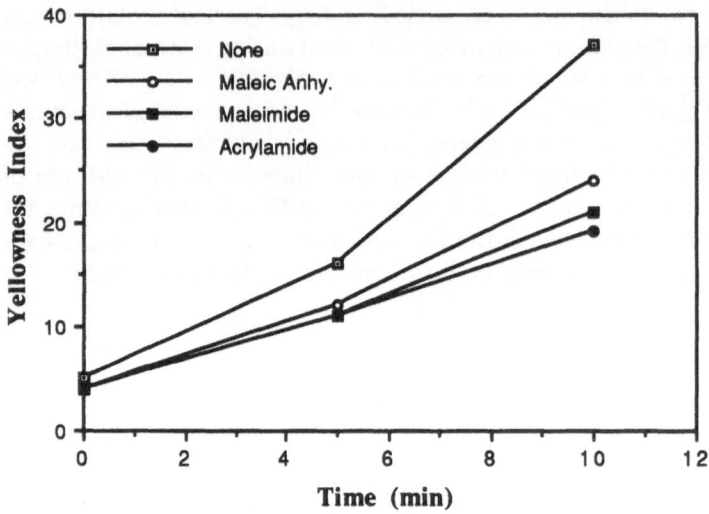

Fig. 23. The effect of dienophiles on the discoloration of SAN [17]

Instead, Marien proposes that linear polyimines form from pendant nitrile groups on neighboring chains which subsequently form intensely colored nitrones upon oxidation. He believes that dienophiles can intercept the polyimine intermediate and disrupt its conjugation (Scheme 8). However, the use of dienophiles to control discoloration is inefficient because the discoloration that does take place is not reversed by the addition of more dienophile. Apparently once the conjugated polyimine becomes oxidized to the nitrone, it becomes deactivated as a Diels–Alder diene. A finding that tends to support the interchain reaction of nitriles is that the solubility of SAN decreases during heating [87].

4 Contribution of Small Molecule Chemistry to SAN Discoloration

A key consideration is the potential for discoloration chemistry of small molecules (monomers and oligomers) to be much different than the polymer, since the oligomers cannot contain long AN sequences. Schellenberg et al., have studied the effects of SAN oligomers on the properties of SAN copolymers. The oligomers were obtained by two techniques: synthesis by inhibited copolymerization of SAN near its azeotropic composition [10], and isolation by partial condensation of vapors during the removal of unreacted monomers from continuously polymerized SAN [9]. Addition of oligomers to SAN has a detrimental effect on color stability (Fig. 24) while improving the balance of flow and toughness.

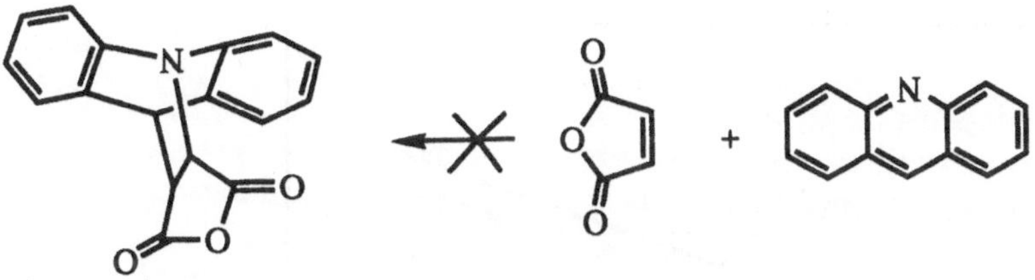

Scheme 7. Attempted reaction of maleic anhydride with naphthpyidine

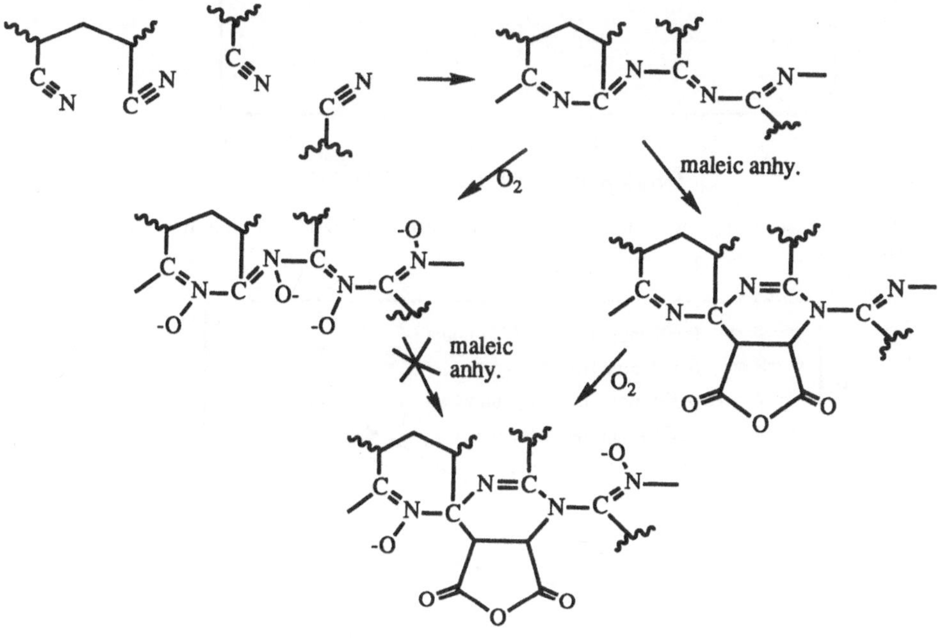

Scheme 8. Interception of polyimine intermediate by maleic anhydride

Allan et al. [67] determined the relative contribution of small molecule and high polymer backbone chromophores by molding two SAN resins (0.395 and 0.492 mole fraction AN) at low (200 °C) and high (280 °C) temperature. The molded plaques were dissolved (10 w/w) in methylene chloride and the color of the solutions measured. Then the high polymer fraction was precipitated (pct) by the addition of two parts of methanol. The pct polymers were dried in a vacuum oven at 60 °C, redissolved in methylene chloride, and the color remeasured. About half of the discoloration was found to be due to small (methanol soluble) molecules (Fig. 25).

The structures of SAN oligomers (Fig. 26) and their mechanisms of formation (Scheme 9) have been studied in detail [88, 89].

During a study of the mechanism of formation of these trimers they were heated to gain further insight into their mechanism of formation [88]. That

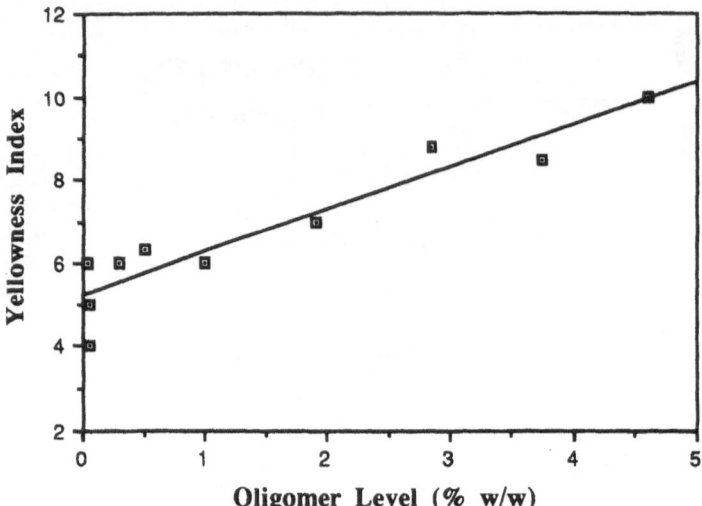

Fig. 24. Effect of oligomer content on the color stability of SAN (0.4 mole fraction AN) during fabrication

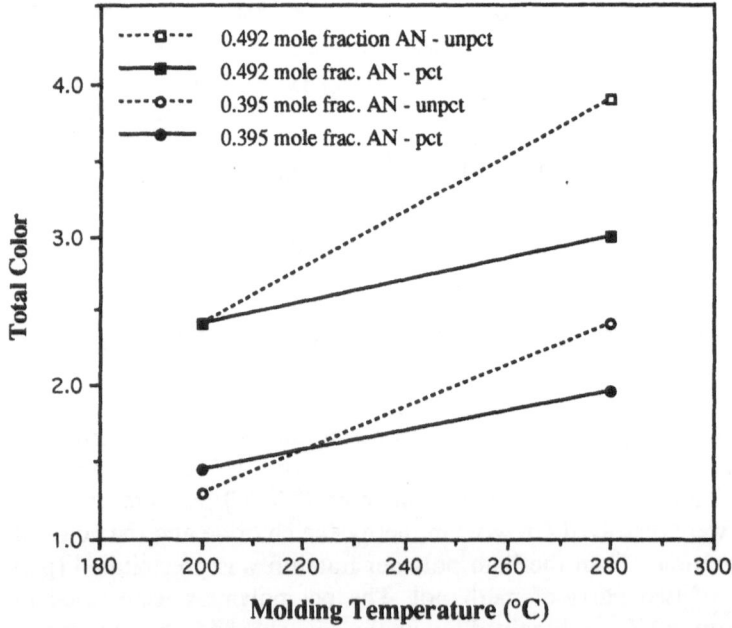

Fig. 25. Color of two (pct. and unpct.) SAN resins (0.395 and 0.492 mole fraction AN) molded at low and high temperatures [67]

experiment resulted in the discovery of a novel decomposition reaction of (THNA) producing a highly UV absorbing and fluorescing 1,2,3-trisubstituted naphthalene derivative [2-amino-3-methyl-1-naphthalenecarbonitrile (AMNC)] (Scheme 10) [90]. They also studied the kinetics (Fig. 27) and mechanism (Scheme 11) of the decomposition reaction of THNA to form AMNC.

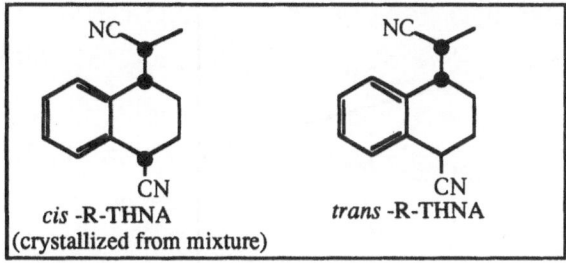

Fig. 26. Structures of the A_2S trimer diastereomer pairs formed during the spontaneous copolymerization of styrene and acrylonitrile

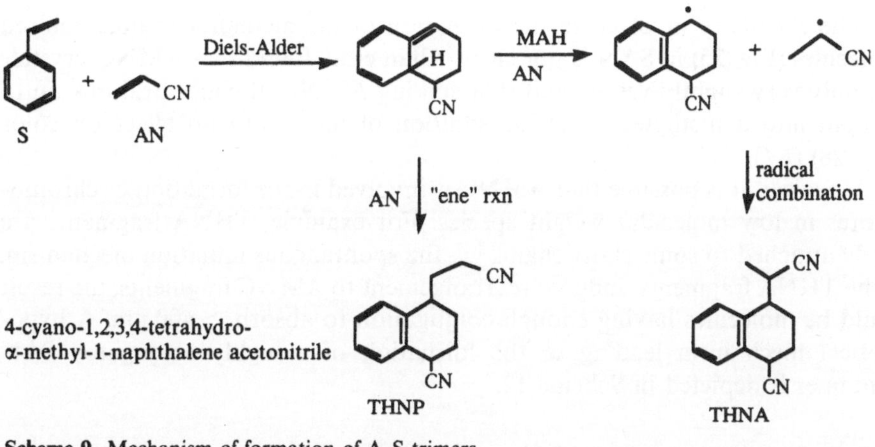

Scheme 9. Mechanism of formation of A_2S trimers

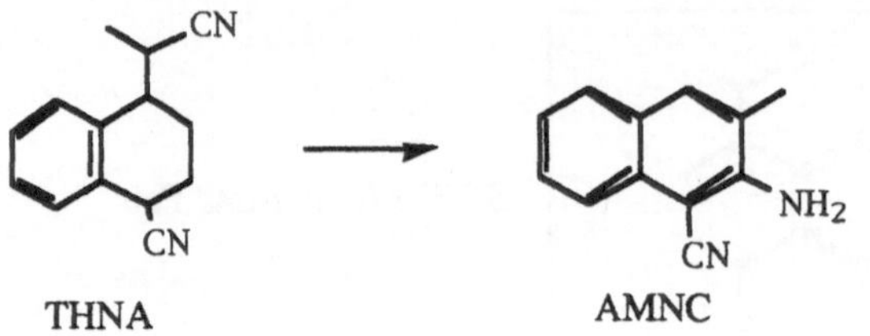

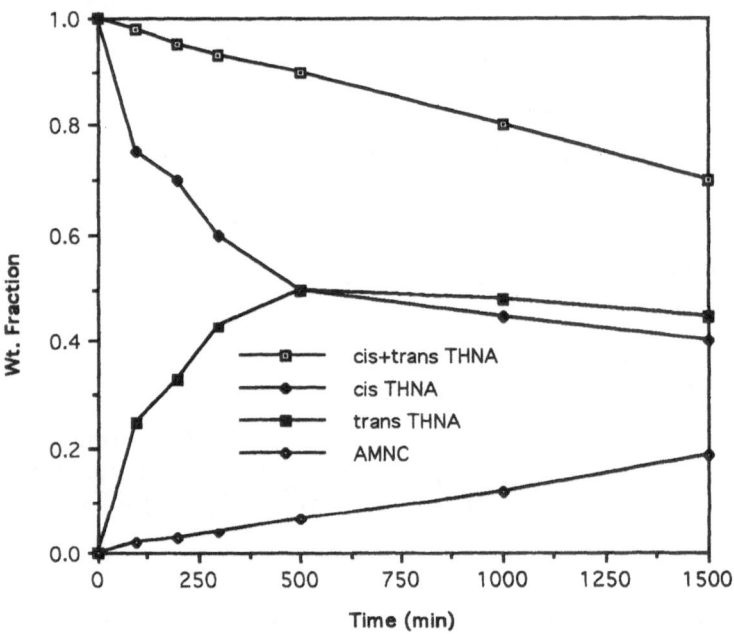

Scheme 10. Formation of AMNC during thermal decomposition of THNA

Fig. 27. Epimerization and rearrangement kinetics of *cis*-R-THNA at 280 °C

The chemistry involved in the formation of the methanol soluble colored molecules (Fig. 25) in SAN is not clear. Allan et al. found that AMNC crystals are only very slightly yellow and that spiking AMNC at concentrations up to 50 ppm into a methylene chloride solution of SAN with no affect on color (Fig. 28) [67].

However, it is possible that AMNC is involved in the formation of chromophores in low molecular weight species. For example, THNA fragments are likely attached to some short chains via the spontaneous initiation mechanism. If the THNA fragments undergo rearrangement to AMNC fragments, the result could be molecules having enough conjugation to absorb >400 nm. A hypothetical mechanism leading to the formation of a highly conjugated SAN pentamer is depicted in Scheme 12.

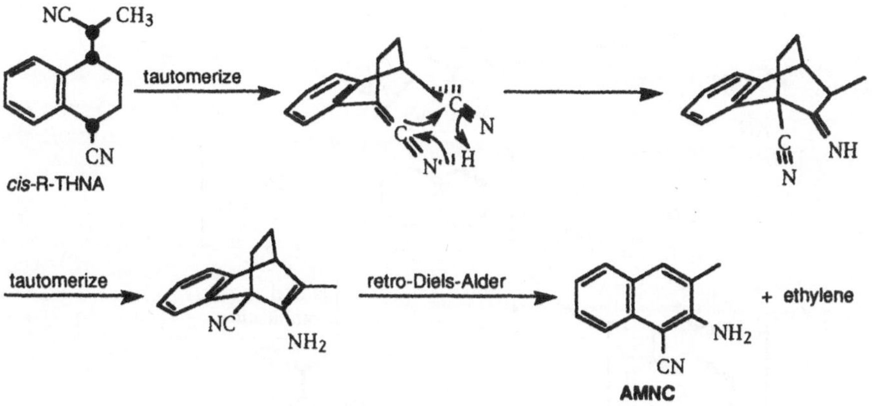

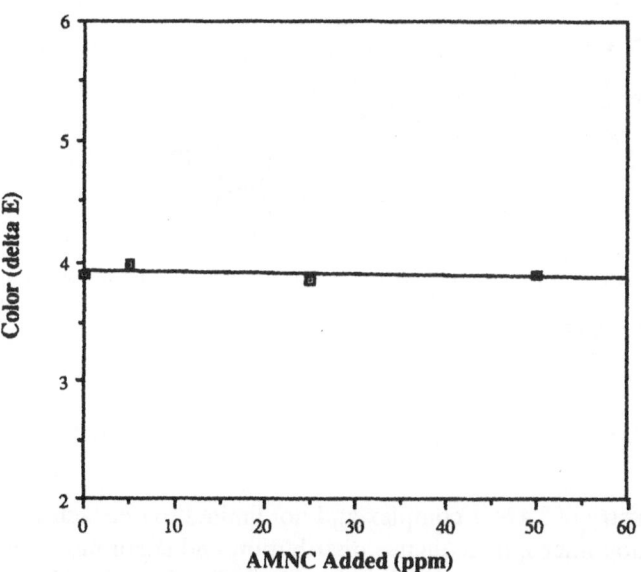

Scheme 11. Proposed mechanism for AMNC formation

Fig. 28. Contribution of AMNC to the color (ΔE) of SAN in solution [67]

Another possible mechanism for AMNC to become involved in the formation of molecules that absorb at >400 nm is by condensation of its amino group with carbonyl containing impurities in the polymer. For example, most styrenic polymers contain traces of benzaldehyde. Indeed, Allan et al. found that heating AMNC with benzaldehyde forms intensely yellow mixtures [67].

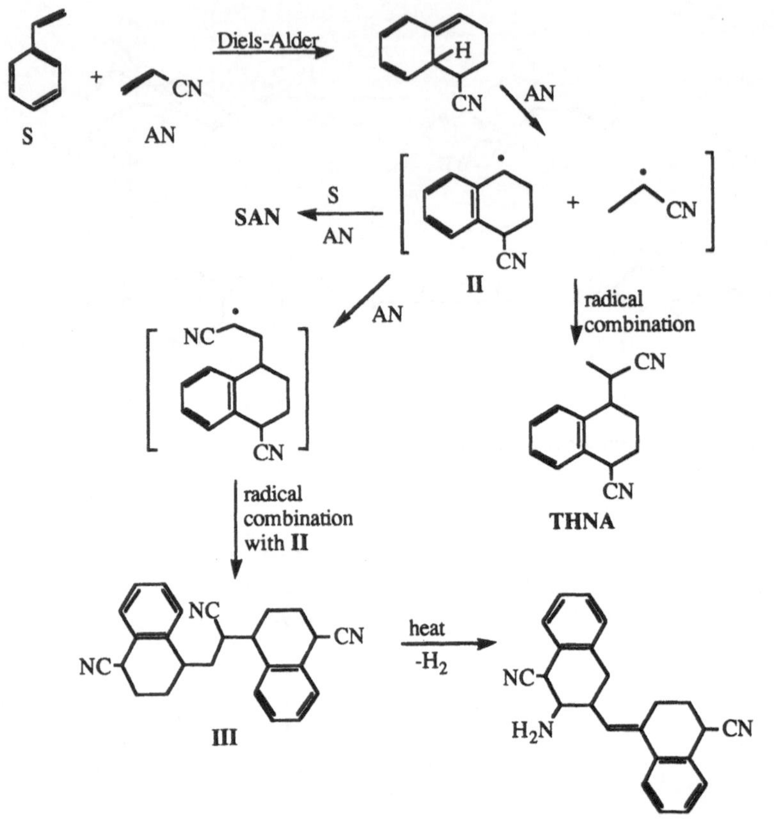

Scheme 12. Spontaneous initiation mechanism for SAN copolymerization showing hypothetical formation of a SAN pentamer containing a THNA fragment which subsequently decomposes to a highly conjugated molecule upon heating [67]

5 Conclusions

The discoloration chemistry of SAN is complex and not limited to one factor or mechanism. Both the monomer unit sequence distribution and small molecule residues in the polymer contribute to the discoloration. The SAN backbone chromophores appear to be located primarily on the chain-ends and a radical chain-scission mechanism involving nitrile cyclization may be responsible for its formation.

The mechanism of formation of chromophores in the oligomeric fraction of the polymer is uncertain. It is possible that oligomers decompose to form chromophores or that the polymer unzips or slits off colored oligomers from the high polymer.

Since it appears that the formation of chromophores in the polymer backbone involves the cyclization of AN sequences, the color stability of SAN should be improved by increasing the alternation tendancy of the polymerization. This is accomplished by conducting the polymerization to minimize the value r_A. The value of r_A can be decreased by polymerizing at lower temperature, or by adding Lewis acid to the feed.

The higher than expected r_A value for commercial SAN polymers made using continuous bulk polymerization at 150–160 °C is puzzling. The discoloration resistance of commercial SAN resins could likely be improved if the cause of the high r_A could be found and the copolymerization process operated to increase the alternation tendency of the copolymerization.

6 References

1. Hanson AW, Zimmerman RL (1957) Ind Eng Chem 49: 1803
2. Hanson AW, Best JS (1961) U.S. Patent No. 2989517
3. Molau GE (1965) J Polym Sci, Part B 3: 1007
4. Illig SA (1989) SPE ANTEC 876
5. Illig SA (1990) Rev Plast Mod 60: 409
6. Yabumoto S, Ishii K, Kawamori M, Arita K, Yano H (1969) J Polym Sci, Part A 7: 1683
7. Yabumoto S, Ishii K, Arita K (1969) J Polym Sci, Part A 7: 1577
8. Yamamoto Y, Tsuge S, Takeuchi T (1972) Kobunshi Kagaku 29: 407
9. Schellenberg J, Hamann B (1991) Angew Makromol Chem 187: 123
10. Schellenberg J, Wigand G (1992) Makromol Chem 193: 3063
11. Wild H, Jung R, Echte A, Zizisperger J, Gausepohl H (1977) US Patent No 4061858
12. Patron L, Bastianelli U (1974) Appl Polym Symp 25: 105
13. Tirrel M, Gromley K (1981) Chem Eng Sci 36: 367
14. Kent RWJ (1981) U.S. Patent No. 4243781
15. Florjanczyk Z, Siudakiewicz M (1986) J Polym Sci, Polym Chem 24: 1849
16. Kent RWJ (1981) U.S. Patent No. 4268652
17. Marien BA (1979) J Polym Sci, Polym Chem Ed 17: 425
18. Johnson J, Potter W, Rose P, Scott G (1972) Br Polym J 4: 527
19. Arita K, Ohtomo T, Tsurumi Y (1981) J Polym Sci, Polym Lett Ed 19: 211
20. Alexandru L, Somersall AC (1977) J Polym Sci, Polym Chem Ed 15: 2013
21. Atiqullah M, Hassan MM, Beg SA (1992) J Appl Polym Sci 46: 879
22. Adama R (1979) Plaste Kautsch 26: 331
23. Johnston NW (1976) J Macromol Sci -Rev Macromol Chem C14: 215
24. Goldfinger G, Steidlitz M (1948) J Polym Sci 3: 786
25. Thomson BR, Raines RH (1959) J Polym Sci 41: 265
26. Chapiro A, Jaberg F, Perec-Spritzer L (1975) E Polym J 11: 637
27. Nagaya T, Sugimura Y, Tsuge S (1980) Macromolecules 13: 353
28. Lewis FM, Mayo FR, Hulse WF (1945) J Am Chem Soc 67: 1701
29. Doak KW (1950) J Am Chem Soc 72: 4681
30. Pichot C, Zaganiaris E, Guyot A (1975) J Polym Sci, Polym Symp 52: 55
31. Pichot C, Pham QT (1979) Makromol Chem 180: 2359
32. Hatate Y, Hano T, Miyata T, Nakashio F, Sakai W (1971) Kagaku Kogaku 35: 903
33. Sandner B, Loth E (1976) Faserforsch Textiltech 27: 571
34. Asakura JI, Yoshihara M, Matsubara Y, Maeshima T (1981) J Macromol Sci, Chem A15: 1473
35. Fordyce RG, Chapin EC (1947) J Am Chem Soc 69: 581
36. Doak KW, Deahl MA, Christmas IH (1960) Abstr Papers 137th ACS, Cleveland, OH 151
37. O'Driscoll KF (1969) J Macromol Sci-Chem A3: 307
38. Uchida H, Aotci Y, Kato T (1993) Kobunshi Ronbunshu 50: 941
39. Schaefer J (1971) Macromolecules 4: 107

40. Hill DJT, O'Donnell JH, O'Sullivan PW (1982) Macromolecules 15: 960
41. Stejskal EO, Schaefer J (1974) Macromolecules 7: 14
42. Sandner B, Keller F, Roth H (1975) Faserforsch Textiltech 26: 278
43. Barron PF, Hill DJT, O'Donnell JH, O'Sullivan PW (1984) Macromolecules 17: 1967
44. Keller F, Sandner B (1976) Faserforsch Textiltech 27: 45
45. Sargent M, Koenig JL, Maecker NL (1991) Appl Spectrosc 45: 1726
46. Oi N, Moriguchi K (1974) Bunseki Kagaku 23: 798
47. Littke WH, Fieber W, Schmolke R, Kimmer W (1975) Faserforsch Textiltech 26: 503
48. Nyquist RA (1987) Applied Spectroscopy 41: 797
49. Wolfram LE, Grasselli JG, Koenig JL (1974) Applied Polym Symp 25: 27
50. Garcia RLH, Ro N (1985) Can J Chem 63: 253
51. Bruessau RJ, Stein DJ (1970) Angew Makromol Chem 12: 59
52. Garcia RLH, Hamielec AE, MacGregor JF (1982) ACS Symp Ser 197: 151
53. Blazso M, Varhegyi G, Jakab E (1980) J Anal Appl Pyrolysis 2: 177
54. Tsuge S, Kobayashi T, Sugimura Y, Nagaya T, Takeuchi T (1979) Macromolecules 12: 988
55. Tsuge S, Sugimura Y, Kobayashi T, Nagaya T (1981) Polym Prepr Am Chem Soc, Div Polym Chem 22: 284
56. Hill DJT, Lewis DA, O'Donnell JH, O'Sullivan PW, Pomery PJ (1982) Eur Polym J 18: 75
57. Bartick EG (1979) J Chromatogr Sci 17: 336
58. Mori S, Wada A, Kaneuchi F, Ikeda A, Watanabe M, Mochizuki K (1982) J Chromatogr 246: 215
59. Ham GE (1954) J Polym Sci 14: 87
60. Guyot A, Guillot J (1967) J Macromol Sci-Chem A1: 793
61. Guyot A, Guillot J (1968) J Macromol Sci-Chem A2: 889
62. Hill DJT, O'Donnell JJ, O'Sullivan PW (1982) Prog Polym Sci 8: 215
63. Sandner B, Loth E (1976) Faserforsch Textiltech 12: 633
64. Harwood HJ (1987) Makromol Chem, Macromol Symp 10/11: 331
65. Hill DJT, Lang AP, O'Donnell JH, O'Sullivan PW (1989) Eur Polym J 25: 911
66. Hill DJT, Lang AP, O'Donnell JH (1991) Eur Polym J 27: 765
67. Allan D, Birchmeier M, Pribish J, Priddy D, Smith P, Hermans C (1993) Macromolecules 26: 6068
68. Yabumoto S, Ishii K, Arita K (1970) J Polym Sci, Part A-1 8: 295
69. Patnaik BK, Takahashi A, Gaylord NG (1970) J Macromol Sci, Chem 4: 143
70. Gaylord NG, Patnaik B (1970) J Polym Sci, Part B 8: 401
71. Gaylord NG, Dixit SS, Patnaik BK (1971) J Polym Sci, Part B 9: 927
72. Gaylord NG, Dixit SS (1971) J Polym Sci, Part B 9: 823
73. Gaylord NG, Dixit SS, Maiti S, Patnaik BK (1972) J Macromol Sci, Chem 6: 1495
74. Gaylord NG (1975) U.S. Patent No. 3919182
75. Gaylord NG, Tomono T (1975) J Polym Sci, Polym Lett Ed 13: 697
76. Seymour RB, Stahl GA, Garner DP, Knapp RD (1976) Polym Prepr, Am Chem Soc, Div Polym Chem 17: 216
77. Sargent M, Koenig JL, Maecker NL (1993) Polym Degr & Stabil 39: 309
78. Sargent M, Koenig JL, Maecker NL (1993) Polym Degr & Stabil 39: 355
79. Mork CO, Priddy DB (1992) SPE ANTEC Proceedings
80. Mork CO, Priddy DP (1992) J Appl Polym Sci 45: 435
81. Warner SL, Howell BA, Smith PB, Dais VA, Priddy DB (1992) J Appl Polym Sci 45: 461
82. Grassie N. MaGuchan R (1972) Eur Polym J 8: 243
83. Grassie N, Hay JN (1962) J Polym Sci 56: 189
84. Grassie N, Bain DR (1970) J Polym Sci, Part A 8: 2653
85. Vukovic R, Kuresevic V, Gnjatovic V (1974) Hem Ind 28: 565
86. Runge J, Nelles W (1970) Faserforsch Textiltech 21: 105
87. Gupta MC, Nambiar J (1983) Colloid Polym Sci 261: 709
88. Hasha DL, Priddy DB, Rudolf PR, Stark EJ, De Pooter M, Van Damme F (1992) Macromolecules 25: 3046
89. Kirchner K, Schlapkohl H (1976) Makromol Chem 177: 2031
90. Stark E, Bell BM, Hasha DL, Priddy DB, Skelly NE, Yurga LJ (1992) J Macromol Sci, Macromol Reports A29: 1

Editor: Prof. J.L. Koenig;
Received 20. Dec. 1993

Author Index Volumes 101-121

Subject Index

Springer-Verlag
and the Environment

We at Springer-Verlag firmly believe that an international science publisher has a special obligation to the environment, and our corporate policies consistently reflect this conviction.

We also expect our business partners – paper mills, printers, packaging manufacturers, etc. – to commit themselves to using environmentally friendly materials and production processes.

The paper in this book is made from low- or no-chlorine pulp and is acid free, in conformance with international standards for paper permanency.